Moon Landing

Learn About the Origins of the Spacecraft

(An Impartial Investigation Into the Apollo Moon Landings)

Kenneth Wilkinson

Published By **Oliver Leish**

Kenneth Wilkinson

All Rights Reserved

Moon Landing: Learn About the Origins of the Spacecraft (An Impartial Investigation Into the Apollo Moon Landings)

ISBN 978-0-9938088-6-9

No part of this guidebook shall be reproduced in any form without permission in writing from the publisher except in the case of brief quotations embodied in critical articles or reviews.

Legal & Disclaimer

The information contained in this book is not designed to replace or take the place of any form of medicine or professional medical advice. The information in this book has been provided for educational & entertainment purposes only.

The information contained in this book has been compiled from sources deemed reliable, and it is accurate to the best of the Author's knowledge; however, the Author cannot guarantee its accuracy and validity and cannot be held liable for any errors or omissions. Changes are periodically made to this book. You must consult your doctor or get professional medical advice before using any of the suggested remedies, techniques, or information in this book.

Upon using the information contained in this book, you agree to hold harmless the Author from and against any damages, costs, and expenses, including any legal fees potentially resulting from the application of any of the information provided by this guide. This disclaimer applies to any damages or injury caused by the use and application, whether directly or indirectly, of any advice or information presented, whether for breach of contract, tort, negligence, personal injury, criminal intent, or under any other cause of action.

You agree to accept all risks of using the information presented inside this book. You need to consult a professional medical practitioner in order to ensure you are both able and healthy enough to participate in this program.

Table Of Contents

Chapter 1: The Moon 1

Chapter 2: The Apollo Program 11

Chapter 3: The Moon Landing Conspiracy 22

Chapter 4: Debunking The Conspiracy Theory On Moon Landing 62

Chapter 5: Moon Shadows 77

Chapter 6: Moon Dust Is Not Present On The Landing Module's Feet 89

Chapter 7: Close-Up Of The Cross-Hair That Is Disappearing 97

Chapter 8: Conspiracy True? 105

Chapter 9: Occam's Razor 116

Chapter 10: Easily-Overlooked Conveniences 124

Chapter 11: The Patriots 133

Chapter 12: What Are Conspiracy Theories? 143

Chapter 13: The Cold War 160

Chapter 14: The Space Race.................. 174

Chapter 1: The Moon

Before tackling the complicated issues of determining whether something is true or false, it's crucial to understand the physiology of our Moon and the physiology of it.

The moon of our planet is the sole satellite that is visible from Earth. It's a chunk of rock that is much smaller than our own planet, and like ours, is an uninhabited planet. It is uninhabited as there isn't any either air nor water. The lunar surface Moon is covered in large craters. They are formed by chunks of rock called meteorites that fall from space. If you're in a clear, dark night, it is possible to be able to observe what's known as "splash" marks surrounding one moon's craters with out having to use an instrument like a telescope.

The Moon is thought to be older than the Earth. It orbits the Earth in a manner that we don't glimpse the other part of the Moon so we are able to see exactly the same part of the Moon each night. The Moon appears to alter its form from night to the following night. This is because, when it circles the Earth (taking around 27 days to complete its circle) it is the same face of it is visible to our faces all day long. This is because it's our perspective of the sun's light upon it that shifts. If the face of ours is turned completely away from the Sun and we are unable to be able to see the Moon even a bit. When the Moon is fully turned toward the Sun and we are able to observe an Full Moon.

Photograph: On the 9th of January 1999, NASA announced the prime crew for the Apollo 11 lunar landing mission. The photo of three astronauts was snapped following the announcement, on January 20, 1969.

FEATURES OF THE MOON

1. The Moon is a smoky ball of rock with a diameter of 376 kilometers in size which is roughly one quarter the size on the Earth.

2. The moon's surface Moon is extremely uneven thanks to huge craters, massive

mountains and flat planes known as 'seas composed of lava that has been hardened.

3. Like we said previously, as we've mentioned, the Moon is Earth's one and only natural satellite, an astronomical body orbiting an earth.

4. The orbit of the Moon around the Earth is designed as a slightly squashed circular shape that is referred to as an ellipse.

5. The Moon takes 27.3 days to allow the Moon to make its distance around Earth and finish its orbit.

6. Even though, the Moon sparkles brightly through the night, it does not generate its own luminescence. The Moon is visible to us Moon as it reflect sunlight that comes from the Sun.

7. Have you noticed that the Moon seems to alter its in shape every evening? This is because, during the time that the Moon moves around the Earth and moves around

the Earth, the Sun shines brightly on different regions of the Moon's surface. So it's only our perception of the Moon and not the Moon itself.

8. The temperature of the Moon can vary from extremely hot to extremely cold! As the Sun is at its highest point the temperatures can rise to the scorching temperature of 127 degrees Celsius. When the sun sets it can bring temperatures down to a low of around -153 degrees Celsius.

9. Similar to Earth Like Earth, Moon is also like Earth. Moon is also a planet with gravitational force (the force that draws things toward the earth). The Moon's gravity however is less powerful, being one-sixth the gravity of Earth. It means you'd be weighing less when you sit in the Moon!

10. Scientists don't know for certain what caused the Moon was created. One popular hypothesis is that a large rock of Mars known as Theia was thrown into Earth

approximately 4.5 billion years in the past. The fragments of the collision then clumped into our Moon.

11. Although Moon's Gravity is significantly less than that of Earth Gravitational forces applied to the Earth through the Moon resulted in extremely long waves which move across oceans. The waves that move across the ocean are known as tides.

Image: Apollo 13 Saturn V rocket engine during transport to the dock for rocket launch.

MOON MISSIONS

Everyone at least once has considered going to the Moon. The dream of mankind to go to the Moon has finally been realized with the Apollo 11 mission. The mission was completed in July 1969. US NASA astronaut Neil Armstrong became the first human to step foot upon the Moon. Neil Armstrong traveled to the Moon together with two fellow astronauts. From 1969 to 1972 12 other astronauts also were also on the Moon. They walked around the lunar lunar surface, captured photographs and conducted important experiments. They also returned rocks for research. The later missions cars-sized Lunar Rover was used to travel and cover longer distances across the Moon. Spacecraft astronauts are secured on the Moon by spacesuits. The suits were equipped with the ability to breathe and regulate temperature along with a radio link to the spacecraft. There were several successful missions completed in subsequent periods by various countries, however we'll focus on the Apollo Space

Program - the initial and final program which made it possible for at least 12 humans to be able to walk through the

moon. Moon.

Photograph: On the 16th of July 1969, 1969 was the historical moment in the history of Apollo 11 with three astronauts being lifted off the ground.

NATIONAL AERONAUTICS AND SPACE ADMINISTRATION (NASA)

Discussions about the Moon Landing conspiracy without talking about NASA is like talking about tea and not drinking water. It is crucial to be aware of NASA. It is important to know that the Apollo Program was funded by NASA.

NASA is an autonomous space agency that is part of the US federal Government located at Washington DC. It was created in 1958 NASA in the year 1958, to assist with the execution of numerous Aeronautics, space-related programs and research in space. The moon missions became possible due to NASA's Apollo Space Program and NASA served as the foundation for these programmes. NASA is a large organization which has its own branch and subdivisions with different goals. A very obscure branch has been believed to exist within a highly restricted zone located in the Nevada Desert is called the "Area - 51".

Area 51 is an extremely classified Air Force facility centre which is situated within the Nevada Test and Training Range. The facility is so tightly guarded that it is not known what the motive behind it is. According to speculation there is a possibility that NASA's scientists are also to create modern warfighter planes as well as weapons. A lot of people have claimed to have seen mysterious flying objects hovering in the area that was most secure. A lot of conspiracy theorists believe that those in the Apollo Moon Landing series were filming in Area 51. The subject will be discussed further in the book.

"Maybe they have all the answers for all the good reasons"!

Chapter 2: The Apollo Program

It was the Apollo Mission is not just the title of a moon-related mission, but rather a sequence of missions which first study the space journey, then study the safety requirements to make a trip to the moon, as well as many other explorations of science connected with the Moon. It was known as the Apollo program, also referred to by the name Project Apollo, was the third United States human spaceflight program that was carried out through NASA. National Aeronautics and Space Administration (NASA) that successful in landing the first human beings on the moon's surface Moon between 1968 and 1972. The Apollo program was originally conceived under Dwight. Eisenhower's time as a spacecraft that could carry three persons in order to succeed the one-person Project Mercury, which put the first Americans in space. Apollo was then dedicated to President John Kennedy. Kennedy's goal in the 1960s "landing a man on the Moon and returning him safely to the Earth" in the speech in Congress on May 25th in 1961. Apollo was the 3rd US astronaut to launch, followed by two individuals in Project Gemini conceived in 1961 to expand spaceflight capabilities in the direction of Apollo.

The Apollo program was created to take humans to the Moon and then safely home to Earth. Six of the Apollo missions (Apollos 11 12, 14, 15 17, 16, and 17) accomplished this objective. Apollos 7 as well as 9 included Earth orbiting missions designed to test their Command and Lunar Modules and were not able to return lunar data. Apollos 8 as well as 10 simulated different components as they orbited the Moon and brought back photographs of the moon's surface. Apollo 13 did not land on the Moon due to an issue however, it also took photos. The sixth spacecraft that was able to land on the Moon provided a wealth of research data as well as nearly 400 kilograms of lunar sample. Tests included soil mechanics meteoroids, seismics thermal circulation, lunar ranging magnetic fields, as well as the solar wind experiment.

It's worth noting in this context that the period of time was one of space races due to cold war that fought between both the Soviet Union (USSR) and the United States of America (USA) and a battle to win the reputation of being superior in capacity for spaceflight. The Sputnik launch was the first launch in 1957, and concluded in July 1969 when the launch of Apollo 11 to the Moon. This is why the Apollo program is essential to comprehend.

Apollo 1 - In 1967 Apollo 1 was scheduled to launch on February 21 But disaster struck just prior to launch. The fire that erupted inside the main capsule the spacecraft, during the pre-flight exercise January 27, 1967 killed all three astronauts scheduled to be on the mission. Virgil Grissom, Edward White and Roger Chaffee all died in the fire.

Apollo 2 and Apollo 3 The Apollo 2 and 3 numbers have not been associated with any Apollo mission.

Apollo 4 Apollo 4 - On the 9th November 1967, first flight of the massive Saturn V rocket launched that was to carry the following Apollo missions. There was no astronaut on the mission.

Apollo 5 - This was another mission that was uncrewed and which was launched on January 22, 1968. It was the first mission of a new space capsule with astronauts aboard in future missions. Spacecrafts of the past were known as crew modules. They were redesigned to be moon module. The older model of the Saturn V rocket, called Saturn IB, was used.

Apollo 6 - On 4th April 1968, the third mission of the Saturn V rocket was launched with no crew. Two engines experienced premature shutdowns as well as a third was not functioning properly however, the mission was declared to be a success.

Apollo 7 - 11 October 1968, it was Apollo 7 was the initial Apollo mission went above the Earth orbit and further to the space. The Apollo 7 crew of Walter Schirra Jr, Walter Cunningham and Donn Eisele was in space for 11 days as they carried out a number of spacecraft-related experiments. The spacecraft was the first to carry out a live TV broadcast. It was a Saturn S-IVB rocket was used.

Apollo 8 - on the 21st of December, 1968 it was the first Apollo mission to circumnavigate the Moon. Astronauts Frank Borman, William Anders and James Lovell Jr also became the first astronauts aboard the Saturn V rocket and spent twenty hours orbiting around the Moon.

Apollo 9 - This mission began on March 3, 1969. The Apollo flight was also the very first mission that saw lunar modules separated from the control module and flew on its own for 6 hours before docking once more. However, the mission was accomplished on the Earth orbit. The crew consisted of James McDivitt, Russel Schweickart and David Scott. Schweickart did a spacewalk.

Apollo 10 - This mission was launched on the 18th of May 1969. It was an act of practice for the eventual landing by humans and was the very first time that the lunar module into the orbit of the Moon. Spacecrafts traveled close to 20 km from the lunar surface before flying across the area where Apollo 11 would land. There were three astronauts on board: Thomas Stafford, Eugene Cernan and John Young.

Apollo 11 - Finally, on July 16, 1969, the journey towards the Moon began. It was the first time man landed on the surface of the Moon. Armstrong and Aldrin wandered around the moon's surface Moon for two and a half hours. They gathered soil and rock samples, conducted tests, and planted the American flag. They stayed for 21 hours, 36 mins in the lunar surface mostly in moon's lunar module. The third member of the team, Michael Collins, remained inside the command module within the lunar orbit. They all returned safe on the 24th of July.

Apollo 12 - On November 14th, 1969, the Apollo 12 mission was launched in order to commemorate the second time that humans landed in the Moon on the 19th of November just six months. It was returned on November. Charles Conrad Jr and Alan Bean took to the lunar surface, while their co-pilot Richard Gordon Jr remained in the command module. The mission returned parts of Surveyor III, a lander mission which had landed on the Moon in the past two years.

Apollo 13 - It was first launched on the 11th of April in 1970. Apollo 13 experienced an error in the flight and the command module sustained damage. The team of James Lovell Jr, Fred Haise Jr as well as John Swigert Jr had to relocate onto their lunar modules. The mission was then recalled and the spacecraft was able to return safely back to Earth.

Apollo 14 - On January 31st 1971 The Apollo 14 crew begun their mission and reached the Moon on the 5th of February. They returned safely to Earth on the 9 February. The Apollo 14 crew was the third person to land in the Moon. The team comprised Alan Shepard, who in 1961 was the first American to go to space. Edgar Mitchell, and Stuart Roosa was within the control module. Shepard took a walk of over 2.5 kilometers on the Moon which was a record for the time.

Apollo 15 - On 26 July 1971, the Apollo 15 spacecraft was launched and made land on the Moon on the 30th of July. The mission was launched again, this time they parked an automobile on the moon's surface, and then drove for 25 kilometers along the moon's surface. Then, they returned to Earth on the 7th of August. The group of David Scott and James Irwin took 18 hours to explore the moon's surface. They also conducted the gravitation test in order to demonstrate the iron hammer, which is heavy, as well as a feather that was light fell at the same time, despite no air drag at the Moon. Alfred Worden was the third astronaut aboard, and was in command of the module.

Apollo 16 - This mission began on the 16th of April 1972. Apollo 16 landed on the Moon on the 20th of April, and then landed on Earth on the 27th of April. John Young and Charles Duke piloted a rover to the Moon and returned over 90 kilograms of lunar material. Third person to fly, Ken Mattingly, carried his spacewalk.

Apollo 17 - On 7th December 1972, Apollo 17's final lunar mission began with the launch of Apollo 17 and landed on Moon in just 4 days, 11th December. The mission was the last of the Apollo program. Astronauts Eugene Cernan and Harrison Schmitt took three days to the Moon and completed three long Moonwalks that lasted seven hours. This mission returned over 240 kilograms of lunar material. Apollo 17 landed on Earth on the 19th of December, 1972.

The total amount of Apollo astronauts brought back more than 350 kilograms of rock as well as soil samples from the moon. NASA reports that it gets around 60 requests each year for the soil and rock samples. A survey conducted by the agency found that more than 2500 scientific papers were composed using data from Apollo

information.

Today, we know that landing humans on the moon was many hours of preparation. If the technology of that time allowed humans to travel to the Moon and back, today's technology, it will be simpler. It's surprising that Apollo was the sole human-powered Moon mission, and nobody from any other nation has ever tried it.

Image: The powerful Apollo 16 rocket engine on being carried.

Chapter 3: The Moon Landing Conspiracy

Conspiracy Theory - "DID WE LAND ON THE MOON?"

Disclaimer: The story is based on a 2001 famous documentary directed by John Moffet. The documentary focuses on an area that has been controversial. The arguments presented in the program aren't an exhaustive list of possible interpretations. The viewers are encouraged to form their own decision on the basis of all available data.

The plotline of Conspiracy Theory

One of the most remarkable accomplishments in the history of mankind has been one of humanity's greatest achievements was the Moon Landing. The day that it happened, 5 million viewers on television watched U.S. astronauts Neil Armstrong and Buzz Aldrin take their first steps onto the Moon. The two astronauts along with the third crew member Michael

Collins flew safely back to Earth and made it to the Pacific Ocean.

Just a couple of years following the amazing feat was announced, a few people speculated that it was a "giant leap for mankind" was a hoax. Theories that moon landings were actually faked by was staged by the U.S. government had staged to beat the Soviets gained traction during the late 1970s. Even though these assertions were debunked and can be easily disproved however, they continue to be popular up to the present day.

A majority of denialists "proof" is based on perceptions of anomalies in images sent back to Earth by the moon's lunar surface. "With few exceptions, the same arguments just keep coming up over and over again," states Rick Fienberg, the press official of the American Astronomical Society, who is a doctoral student in Astronomy. Fienberg has firsthand experience of this. Nearly forty years back, Fienberg debated one of the most prominent moon landing denialists, Bill Kaysing, on TV.

The first claims that the moon landing was fabricated occurred at the time that the Pentagon Papers as well as Watergate have shattered Americans confidence in their elected officials. The idea of faking the accomplishments of Apollo 11 Apollo 11 mission would require an elaborate scheme of deceit that it would almost impossible Fienberg says. Fienberg.

"About 400,000 scientists, engineers, technologists, machinists, electricians, worked on the Apollo program," Fienberg notes. "If there is a primary motive behind belief in the moon hoax is that you don't trust our authorities, you do not trust the leaders we have and you aren't confident in authorities, then how do you imagine that 400,000 people will not speak to each other for over 50 years? This is just not possible."

ENOUGH OF TALKING, LET'S BEGIN!

This article will look into the most remarkable event in the 20th century: astronauts getting on the moon. It's true that there are those who claim the moon landing never took place. Take your own decision as we look into the facts by analyzing official government images look at the film footage, as well as hear from one former astronaut who's shy to voice his opinions. Did the government orchestrate the largest deception of our time? The jury is out in "Conspiracy Theory - Did we land on the moon" ?

The 16th of July, 1969, America held its breath. Apollo 11 blasted into space starting its journey of 250,000 miles towards the moon. In their 8-day journey they Apollo 11 astronauts saw spectacular landscapes, suspended in the air without weight and went to places that there was no one else before.

Did they actually get to the moon? We all think they did. Many millions of viewers watched on the television when the lunar lander landed and those unforgettable words were said in the lunar lander's landing "At One Small Step for Man, One Giant leap for Mankind - by Niel Armstrong". However, even now people have said that believing in the one tiny step of man requires only one huge jump of faith.

Bill Kaysing was an analyst and engineer for Rocketdyne which was the firm that developed for the Apollo rockets. "There were many problems that evolved during the 60s that led people to believe that we're never going to make it to the moon."

When, in the year 1970, people around the globe watched the Apollo lunar landings Bill Kaysing was watching too. What he saw on TV, in conjunction with the experiences he had as a Rocketdyne turned him into a skeptical. "The entire thing appeared to be fake for me. I believe it was an intuitive sense that what was shown wasn't authentic" Kaysing. Kaysing. After examining the film more deeply and was shocked to discover a number of contradictions. Kaysing noted that regardless of the clearness of space deep, stars weren't visible in the lunar night sky. Kaysing saw the American flag flying even when there was none of atmosphere in the lunar surface. He also discovered there was no blast crater underneath the lunar lander, where its rocket engine was powerful and been fired. The evidence proved Kaysing that we had never sent anyone to the moon. However, NASA denies these claims.

"They're always gonna be people who believe some outlandish theories and the notion that we somehow were able to fake the lunar missions is pretty outlandish - Brian Welch, NASA Spokesman".

However bizarre it may be, it's reported that 20 percent of Americans think that we have never been into space. How can anyone believe it is possible that the most significant moment in history could be an untruth? Does it even make sense that NASA lies to people around the globe?

According to an astronaut who was a part of the Apollo program this is a possibility. "Regarding the Apollo mission, I can't say 100% for sure whether these men walked on the moon" -- Brian O'Leary. Brian O'Leary was an NASA astronaut during the 1960s. He was advisor for science in the Apollo moon mission. "It's possible that NASA could have covered it up just in order to cut corners and to be the first to allegedly go to the moon". Is going to the moon in the first place so vital that our government think about fake moons?

In order to answer this question for the question, we must go back to 40 years ago, the time that America as well as the Soviets were in a battle for dominance. "People believed that the country which was the winner of the race to space would prevail in during the Cold War. The definition of that was being first to moon. This was the time that was characterized by more or less the national panic." Howard Mc. Curdy, PhD, Space Historian, American University. The 4th October of 1957 was the day that the Soviets scared America with the launch of Sputnik as the world's first satellite, into orbit. "The New York Times had to publish an article explaining to Americans that it did not carry nuclear bombs that could be dropped on the city from that altitude". The American citizens' terror of nuclear destruction increased when Russia became the leader in space-related battles.

"Our speaker announced in Congress that we may be headed for extinction" -- Julian Scheer (Former NASA spokesperson). There was a lot of concern that the Soviet Union's main objective was to establish the foundation of a missile station at the lunar surface. While America's space program had a difficult time getting away from the earth. "The probability of reaching the moon and then returning in a safe manner to Earth was similar to 0.0017 percent. That's basically impossible. The thing that really happened to me is that in the 1960s, people would say "if you don't create it, then fake it" Then they would go on to say "fake it" Kaysing.

However, if you believe that the Apollo missions were fakes How was this massive fake created? According to Kaysing that it was the time of launch that the Apollo Saturn five rocket was authentic. The rocket never actually sent spacecraft to moon. "The astronauts were launched by the Saturn five. In order to explain their disappearance it was just that they orbited Earth for 8 days. In the meantime they showed the fake images of astronauts orbiting the moon. On the 8th day, the console was separated from the vehicle, and it fell into earth. This was a sequence that was shown in movies".

The theory was the basis for the film from 1978, "Capricorn One" in which the government is trying to deceive the world by pretending to be on a spacecraft mission to Mars. "We do not claim this planet in the name of America, we claim it in the name of all the people on the planet Earth" film Dialogue. The Apollo footage appears strikingly similar with the footage from Capricorn One. The producer Paul Lazarus suggested the film's storyline may be more real rather than fiction. "I believe that had they desired to do that, NASA could be able to pull off the biggest ever hoax, not sending anyone to the moon, and then replicated the story in a TV studio. In my opinion, it could have been accomplished in the era when technology was there. What we showed in the display was actually our simulation of what we could create within the 4,000008 dollar budget." -- Paul Lazarus.

However, with the budget of $40 billion, Kaysing believes that they were able to make a fake of the moon even if they didn't get into space. "The reason I believe that NASA and the government faked the moon landing was basically it was technically impossible to do it and they simply had to come up with some sort of alternative that they felt the public would believe" Kaysing. Kaysing.

Kaysing believes that lunar landings actually were filmed in Nevada's high desert. This is the secret military base called Area 51. "Area 51 is one of the most tightly guarded areas within the United States. If you tried to enter trying to find details, you may be targeted for shooting and killed at any time without warning" The Associated Press reports that Kaysing.

Russian spy satellite images of the area 51 show not just a set of hangars resembling movie soundstages, but also barren moon-like zones, that happen to be lined by large craters. Take a look at this image of a moon crater, allegedly captured off the moon's orbit Apollo 10. And this satellite picture of the crater in area 51. Astronauts are also aware of the similarities to the surrounding terrain. "It offers a striking aesthetic that is its own. It's similar to the High desert in America." United States" - Astronaut. Can billions of people truly were fooled into believing that the Nevada desert was moon-shaped? Kaysing believes this is possible and could be the reason for why Area 51 is highly protected. "This is a very secret base and with good reason, because undoubtedly the moon SATs are still there" Kaysing. Kaysing. If so, no one is likely to be looking at them any time soon.

It is common knowledge to believe that on the 20th, 1969 the lunar excursion unit is also referred to as the LEM was the vehicle that carried American astronauts on the surface of the moon. But could be used as a prop that was dropped by wires on a set for a movie. Bill Kaysing says that this might be the cause of no engines on the authentic

NASA footage. "The sound level of an engine in a rocket is close in the 140-150 decibels. That is, it's extremely high in volume. It is possible to discern the voices of astronauts on top of the moving engines of a rocket"?

Do you think this is evidence that the footage was authentic? It was filmed in a controlled space this on Earth only a few months prior to the historic landing. The prototype limb was evaluated in Ellington Air Force Base. As NASA cameras film the flight tests, Neil Armstrong struggles to

manage the hefty aircraft. After about 300 feet high, the craft is ablaze and out of control. Then, at the end of the second Armstrong is able to eject and then floats back towards safety. If the lunar lander is so fragile and difficult to maneuver in the controlled atmosphere of Earth How could it be able to LEM fly six times without a glitch within the arid lunar environment?

Photograph Astronaut Neil A. Armstrong, Apollo 11 mission commander, is safely lowered to the ground following a mishap during a rehearsal.

"The LEM had a single engine that was mounted at the center with a couple of thruster jets, push jets and a few of them at the their top. The idea was to regulate their height as they came down. Well, I'll tell you a secret. In the moment you move your tail inside that cabin one inch, you could alter the pattern of load and it would start to pivot and begin spinning" -- Ralph Rene, Author/Scientist.

"The arguments presented by those who believe the lunar landings are a fake are extremely complex and have in support of a belief similar to this. At the end of the day, there's only one piece of evidence that's undisputed and that's the existence of footprints and plans and footprints that remain in the lunar soil" Brian Welch. Brian

Welch.

Image: A boot print is visible on the moon's surface.

The conspiracy theorists claim the footprints themselves appear to be suspicious. "To be able to have a massive rocket engine that blasts at the surface of the moon blasting off all dust, and find footprints that surround the lunar landing pad. To me, that would be impossible for Kaysing.

The photo after the photo shows how the surface of the moon around the LEM is dotted with footprints. However, Kaysing claims that there's something harder to understand. "The fact that there is no blast crater under the LEM is one of the most conclusive pieces of evidence that I find supporting the hoax". There is no evidence of a blast-crater is evident in any of the moon's six lunar landings.

LEM expert Paul Fjeld says he can provide an explanation for why lunar modules didn't leave any crater on arriving at the lunar surface. "The volume of thrust will be required from the lower part of the engine that is used for descent can be anywhere from 1500 to 2000 pounds thrust. It's all it accomplishes is to remove dust". There's nothing burning or similar to that" Paul Fjeld. However, NASA's own illustrations clearly illustrate an explosion the crater.

"And there's another issue if they actually hit the moon the dust would then fallen into the lunar lander in the footpads, and we have no evidence of dust on the feetpads. As I found this out for myself I was like "no there's no way! Was I gazing at the lunar lander, which has is afloat on the moon"! - Kaysing

It could be possible that there was a possibility that LEM was merely a prop in a massive lunar movie setting? "When Armstrong said, that one small step for man, one giant leap for mankind, the footprint that he made; that have easily been made in area 51".

Kaysing says that the LEMs separation from the lunar surface quite a mystery. "In the video of the ascent stage moving upwards, the only thing you observe is a plume of exhaust emanating from the nozzle of the rocket engine. What do we actually see? The ascent stage appears to appear suddenly, without emissions whatsoever, almost as that it had been jerked upwards with an electrical cable".

Do you think this is evidence of a plot? Did the government have the capability of this massive cover-up?

"To propose that this was all faked and a hoax, they have to say that every piece of evidence that every physical scientific test that one could offer to support the reality of the lunar landings, they have to say that all of those are fake" Brian Welch. Brian Welch.

"I would not say that my conclusion that Apollo was not real wasn't based on any one particular element of evidence, but it was a cumulative conclusion. It was all fake" Kaysing. Kaysing.

If moon-landings really were filming on a set for a movie and then, where is the proof? Based on David Percy, an award-winning photographer and filmmaker The proof lies within NASA's lunar images and videos. "Our study suggests that the images from the

Apollo landings do not provide the most accurate or reliable records. According to us, the Apollo images were made up. A lot of them are full of anomalies and consistency" -- David S. Percy.

Actually, Percy claims that when looked at, these pictures suggest that mankind never did go to the moon in any way. The iconic image of man making his first steps onto the lunar surface is of the most well-known ever recorded in the history of mankind. Why are these significant photographs so blurry and difficult to appreciate? NASA states that it's the result of the 1960s' technology.

"But when you look back and examine it you will see that you will see that the Apollo 11 mission was some very poor video, even by our current norms. They were spooky images that didn't look authentic even. This was due to the nature of the radio transmitter in that time and the camera of the day which we could use to allow us for flying during Apollo 11" - Brien Welch.

Investigative Journalist Bart Sibrel believes that NASA deliberately made images difficult to discern. "NASA has orchestrated this hoax using a unique method by using television. The show had one image that they controlled completely both in black and white. It was so which was shaky and convinced everyone that it was a moonstruck show We had no reasons to doubt it. They had total control on the photos, as well as the audio. Sad to admit that it was more simple than most people think" Bart Sibrel (Investigative Journalist). Bart Sibrel (Investigative Journalist).

However, despite the absence of information, conspiracy theories have evidence that suggests these photographs have been staged. While it is apparent that the astronauts are navigating in the lunar gravity that is only one sixth of that of Earth, Percy notes that when the speed of film is doubled, astronauts seem to be moving in Earth's gravitational force. In addition, if the film from the moon rover increased in speed by a factor of two appears as if the rover is actually driving in Earth. However, there's another reason to consider that that the Apollo missions were recorded in Earth.

If there's nothing but breeze or air on the moon, then why do we see the American flag flying? "The Flag is flying on the moon when there is no atmosphere. It implies that there may there was a breeze blowing through the area 51, where they took this" Then they went on to Kaysing.

Are these ambiguous photos simply are the result of astronauts trying to put flags on moon's surface? Perhaps there is more taking place than what is apparent? Do you think still photography is a good idea? Many believe that the style of these bulky spacesuits could make it very difficult for astronauts control their chest-mounted cameras. The person who invented the cameras was Jan Lundberg.

Photograph: This image from the Lunar Module at Tranquillity Base was shot by Neil Armstrong.

"Once you were on the moon at the lunar surface, in the gown and in the Life Support System, you could not view the camera. It was impossible to bend your head so far. The cameras had no viewfinders, they were required to focus with their bodies" The author is Jan Lundberg. If the cameras were that difficult to control, what are the thousands of images captured with clear clarity and perfect framing? We see images appear to be taken from the moon are flawless. If we examine them more closely, Kaysing says flaws begin to appear.

"Unfortunately, errors were made which are now being discovered". Theorists of conspiracy argue that the lack of lighting is a significant issue in lunar photographs. In Kaysing's view was that, when on the moon the only source for lighting is the sun. "They had no extra lighting, no flashes, no things like that". In this image taken by Apollo 14, the shadows have been directed in different directions. This suggests several lighting sources. "The shadows cast by the rocks in the foreground should have been East-West like the LEM's shadow". In this image taken from Apollo 17. The shadows appear to be directed in various directions. "Outside in sunlight shadows always were in parallel with one another, so the shadows will never intersect".

Some conspiracy theorists believe it's just shadows, which suggest that additional lighting was used. What has been documented through the shadows. As an example, below is an astronaut falling into a gigantic shadow cast from the lunar module. His entire body remains evident. What is the reason there is no shrouding of the darkness? The same technique was used on an earlier Apollo mission. The astronaut in this case is clearly lit, and the background is evidently a shadow of darkness. The picture shows that the sun appears directly in front of the astronauts. It is the form of a silhouette. But even the tiniest details of his outfit are instantly recognizable.

"It is as if he's in the spotlight. I'm not able to explain why. I don't know why".

Finally, in this image, with the sun in the background of the moon's module the back of the craft is evident. The text, United States are crisp and clearly visible. Why can these images appear so clear? "It's because there's more than one light source, which means they're not on the moon" Bart Sibrel. Bart Sibrel. However, NASA does not even acknowledge these points.

Photograph Astronaut Edwin E. Aldrin Jr. Pilot of the lunar module exits into the Lunar Module (LM) Eagle and starts to descend.

"There is a myriad of reports that suggest the photos captured of Apollo astronauts were fakes. In fact, there are numerous, it's an ineffective exercise to attempt to find the answers" Brian Welch. Brian Welch.

However, the question remains. What is the reason why some of these photos taken in different time frames, and in different locations, seem to share the same background? Both of these photos appear to share the same mountain background. However, the lunar component can only be seen in one. This seems to be impossible. It is because the LEM was never relocated or changed its location, it remained at its base throughout the entire mission.

There are theories that the same artificial backdrop was used for two distinct images. Background differences are evident in the lunar footage. "The most reliable evidence is peculiarities in the photo document of the journey towards the moon. There's one from Apollo 16, where the identical shot of the same field is seen on two different dates". This video was taken in what was thought to be the very first of the Apollo 16th lunar excursions. The video came from the following day, at another site. NASA said the place was located two and half miles further. However, when one of the videos is overlayed over the other, they appear to be similar. " The conspiracy theorists see that as evidence that we didn't go to the moon but it was staged and the opposite point of view is that it's a case of bad editing". The conspiracy theorists argue that an examination of these photos suggest evidence of a deliberate doctoring.

To be able to refer back Crosshairs were permanently imprinted into lunar cameras, which meant that they'd have to appear at the each image's top. However, in this photograph there is a crosshair behind portion of the lunar Rover. "The situation is impossible and has to be the result of technical manipulation and doctoring are the image". In this image taken from Apollo 11, the equipment that is in front of it covers the crosshair, not hidden in the background. Then, in another photo image from Apollo 12, the American flag covers one crosshair while the astronaut is covered by one of the crosshairs. If presented with these doubtful photographs and videos NASA dispels conspiracy theories.

"Some are extremely complex to hilarious. Arguments that are incorrect optically and physically. They're also wrong they're not scientifically correct, and they're incorrect in the past. There's A lot of claptrap which is kind of interspersed into these arguments" Brian Welch. Brian Welch.

Yet, regardless of what NASA declares that they are a conspiracy theory, some still believe that Apollo was not a fake.

"When I was looking at every single image and footage, I'm certain, and I'm betting my entire life on the fact that we did not go to the moon. I am certain that we weren't.

"How can NASA create such a massive fraud without someone from the inside making the announcement? Virgil Gus Grissom was selected to be one of the initial seven astronauts. He was a father man, and veteran of many space flight. He was an

American hero and is likely to become the first human to step through the lunar surface. However, Grissom was also a loud disdainful of NASA's space program and was reported to have said that somebody would die. But Grissom's biggest fear was quickly realized. In January 27, 1967, just two years after the moon's first lunar landing Grissom and his crew climbed into Apollo Apollo one capsule for a fully-scale simulation. Problems began immediately. The capsule's communications systems stopped working. Then, the spacecraft caught fire with astronauts trapped within. Tragically, Gus Grissom, Ed White and Roger Chaffee perished their lives prior to departing the launch pad. Gus Grisom's relatives believe the Apollo one explosion was not an accidental. "I believe it was deliberately targeted by someone. In my head what was discovered during the investigation into the accident and how by the CIA responsible or who else. It was done with intention" -- Scott Grissom (Son). The family of Grissom doesn't know the person responsible for the death of his son or how this happened. However, they claim NASA has the facts.

Did Gus Grissom on the Apollo one of the astronauts killed in an unfortunate accident or were they silenced due to their knowledge? We will never be able to be able to answer. The reason for the fire remains unsolved and the spacecraft remains in the military base. The cause of the fire is still unknown. Grissom wasn't the sole Apollo critic who suffered an unfortunate and tragic death. Thomas Ronald Barron was a Safety inspector at the time of Apollo one's development. Following the incident, Baron testified before Congress on his belief that the Apollo program that was in confusion, he believed and that United States would never make it to the moon. Baron claimed that his views were a reason to make him the target of.

In the course of his testimonies, Baron submitted a 500 pages of report that outlines his observations. "With real fear that the program could be stopped dead in his tracks" Baron told Julian Scheer. After exactly a week his testimony, the car of Barron was hit by an train. Barron as well as his stepdaughter, and his wife died instantly.

"I believe that Thomas Ronald Barron was murdered. Since he was the only one with an honest story to share regarding Apollo. Apollo Project". Reports of the veterans disappeared mysteriously. To this day there is no evidence to suggest that it was discovered. However, the Apollo program continued, and so did the string of unfortunate deaths. From 1964 to 1967 the total number of astronauts perished through tragic incidents. The deaths of these astronauts represented an astounding 15 percent of NASA's astronaut corp.

If you want to keep an untruth wrapped and hidden from view it's necessary to get rid of anyone who can discuss the matter. Would the government go so far as to carry out this kind of elaborate ploy? NASA declares it impossible.

"There likely was around a quarter million who were directly associated with the Apollo program, and an additional half million who are involved beyond this. A quarter of a million individuals cannot keep their secret in that way. This isn't going be the case".

However, Bart Sibrel insists that most NASA employees did not know about the falsehood. "Very very few in NASA know about this. This is a departmentalized affair. it's the same person that is putting in bolts at Houston or working in Seattle or even in Florida. Nobody knows the complete image. Therefore that no one knows the entire picture aside from a few individuals".

If conspiracy theorists are correct and only a small number of individuals knew all the details and the truth is in the shadows for all time. Evidence of mysterious deaths, falsified photos, and flags waving across the airless void of space aren't just the sole reasons to question the possibility of ever going on a trip to moon. There are those who believe that astronauts may have never survived their journey.

"The reason why they couldn't go to the moon is because of a phenomenon that few people know about called the Van Allen radiation belt". The Van Allen radiation belt is located 500 miles over the Earth These massive radiation bands surround our planet and cover thousands of miles in thickness. "Any human being travelling through the Van Allen belt would have been rendered either extremely ill or actually killed by the radiation within a short time thereafter".

Apart from those of the Apollo missions, there is no human spaceflight ever attempted to travel through the deadly radiation. "Every manned mission in history

Gemini, Mercury, Skylab, the Space Shuttle has been below the radiation belt, all except going to the moon".

In order to protect astronauts the spacecraft will require the protection of six feet of lead. According to the physicist Ralph Rene. "Obviously that the only shield they used was the thin outer foil made of aluminum as well as their suits made up of glass fibres, aluminium fibres, as well as silicon rubber.

"It's fascinating to know that astronauts were shielded with a very thin layer made of aluminum. While on Earth they use the shield of lead on us, when taking an X-ray taken by a dentist".

Many speculate that if Van Allen belts didn't kill the astronauts then more lethal doses of radiation further into space might have. Explosions that are violent in the sun, are known as magnetic storms. They inundate space with a tonne of radioactivity. "Well, the magnetic storm will come along and that can increase the intensity of the radiation belts by maybe 1000 times above what it was before" Geoffrey Reeves, Dr. Geoffrey Reeves.

According to Rene The Apollo 16 mission coincided with one of the sun's largest and most powerful storms that have ever been observed. "Around the rotating sun came this immense flare the biggest one of the 20th century went on for three or four days, all the while it's slowly rotating around". The effects of radiation can be devastating that range from loss of hair to death from cancer The solar flares were absolutely no negative impact upon those who were part of the Apollo 16 crew.

NASA encountered another challenge with respect to the lunar surface is entirely inhospitable space, that is, in the darkness which includes all that is in the shadow of the spacecraft. Temperatures drop to 250 degrees below zero under the solar system, while temperatures rise to up to 250 degrees higher than zero.

Rene was also of the opinion that astronauts' spacesuits that were liquid-cooled would not have offered enough protection against intense radiation and heat. However, NASA affirms that this theory is not true. "The claim that the radiation on the lunar surface would have incapacitated or hurt the astronauts as equal parts, bad science and in a bad understanding of how we went about designing the equipment, the spacesuits that they wore were incredibly tough and very resilient to lots of different things".

"If those suits do what NASA says they can do, then I want to see them; suit up a guy or two put them into Three Mile Island, the pit there that's still hot, and have them clean up the mess but they can't, they don't".

It is a fact that no Apollo astronaut ever had an illness that was serious from a mission towards the moon. Perhaps this is because they were never able to leave the security of the Earth's atmosphere in beginning?

"This is the main reason why the Russian never really intended to send the men to the Moon".

Is it worry about radiation lethal which prompted the Russians to stop going on the Moon? As per one of the chief Cosmonauts they were definitely the cause of their fears. "Of course, we're anxious to venture in the uncharted space. We are, of course, scared. We don't know the way a human being could be in danger from radiation. We thought that the radiation might even get through the structure the craft" Boris Valentinovich Volinov.

Are the dangers that come with space evidence that NASA did not really believe in its Apollo missions? Even to this day, Russians haven't been able to send a person to the Moon up to the point that we don't have plans for returning.

Are you convinced that it is feasible for the government of our country to create such a massive fake? Do you think it's possible it's possible to justify the $40 billion label? It is possible that the Apollo program is nothing more than the most costly film to ever be made.

"It has been my personal conviction that I am 100% convinced of this after many years of study I've conducted on this subject and the fact that NASA did not ever land a person on the moon. Moon.

I'd suggest that anyone who is of the opinion that we didn't go into space is a complete nuts.

Anybody who wants to make me a coup nut, or crack pipe. You're free to do so. However, they're also invited to examine the evidence that is all around.

The bottom line is that The United States went to the moon in the 1960s. It returned later in the 1970s. The end of the story.

Do we have a way to end this debate and put it to rest for the last time? One thing that experts are unanimous on is it's one quarter of a million miles from the moon. If NASA did indeed land on the moon's surface, traces from the six successfully completed Apollo missions would be removed from the base structures of the LEMs and lunar spacecrafts and the lunar rovers. And even the American flags are in place at the landing site.

"I want to encourage NASA and all their loyalists to use the largest telescope available on Earth to see if they can find any lunar landing site. If there's any lunar landing there and I'm not going to say a thing about the Apollo hoax. If there's not a lunar lander in the area, I'll put aside my claim".

However, there is no telescope to study the Moon with such precision. What objects of the Apollo missions linger in the Moon in a silent way or are the conspiracy theories true?

In the next two years Japan will launch an orbiter that will take close-up photos of the moon's surface. What do they expect to discover? In the meantime, the problem is

still if did we get to the Moon?

Photograph: Earth rising from the lunar Horizon.

Chapter 4: Debunking The Conspiracy Theory On Moon Landing

It was a crucial mission, and it was set be a landmark in history. The moon landing was broadcast live on the television. A few people closely watched the Moon landing film footage and images taken by astronauts, and began an investigation. Through careful analysis, they discovered numerous flaws and anomalies which raised serious doubts regarding Apollo programs that brought astronauts onto the Moon. They were convinced of the government's involvement in fakeing the Moon landing, and that's the way in which they came up with the Conspiracy Theory began. The events and situations that trigger suspicions of a conspiracy perpetrated of the US Government in reality required the most in-depth understanding and knowledge.

The momentous event that was Moon Landing was actually a fraud perpetrated of government officials of the US Government to win the space race. Space races with the Soviets began to gain momentum around the middle of the 1970s. However, the claim was utterly false and completely unfounded. Even though it is still in use until today.

Rick Fienberg, the press director of the American Astronomical Society, who has an PhD in astronomy, has rebutted the idea that the scientists and engineers at NASA associated with the Apollo program were unaware of the fictitious Moon Landing. He believes the truth can't be hidden for as long. "About 4000 engineers, scientists technologist, machinists electrical engineers, were part of the Apollo program. If, in truth, the primary reason of believing about the hoax moon is that you don't trust our authorities, you don't believe our leaders, and you aren't sure about authority, then what is the reason the 400,000 would be able to not speak out for fifty years? This is just not possible" Rick Fienberg. Rick Fienberg.

1.The communications system in the spacecraft Apollo 11 stops and fails at some point after launch. What then happens to the astronauts who communicate from their spacecraft the control room located on Earth?

The question has not been answerable. A minor issue can arise at any moment, however it is fixable.

2.The astronauts put the USA flag over the moon's surface that was waving in the sky. If there was none of the moon's atmosphere and the flag is not moving as shown in the official footage that was released by NASA?

According to Fienberg this is because the flag isn't a normal one. If the astronauts would have planted an ordinary flag in the space, it'd have floated in the same way as flags are on Earth even when there's no breeze. It wouldn't be an attractive photo therefore NASA created special flags specifically that astronauts could carry to the of the six Apollo missions.

The flags were hung with an elongated rod in the middle that allowed them to stand to the flagpole. It was discovered that the Apollo 11 astronauts had trouble in extending the rod to the fullest extent when taking still photos the rod results in a ripple that gives the flag the illusion that it's fluttering in the breeze. When you watch video footage that show the flag we can discern that the flag is moving only when the astronauts grind into the lunar surface. Once the astronauts have left they remain at the same bent form because of the partially stretched rod.

3.During daylight the primary source of sunlight came from the sun. It was also observed that shadows emitted by objects such as astronauts and spacecrafts, rocks, etc. at the lunar surface are reflected in various direction. It was evident that there are additional sources of light or else how is it possible?

Shadows created by any two objects are together, however only if the surface is uniform. The lunar surface is completely uneven, with numerous the craters of highlands as well as a few. It is therefore possible that shadows created on the lunar surface from a different direction.

4.The images of the moon's surface appeared extremely similar to the American Area 51. It is heavily guarded by American military, though there's an absence of any significant establishments in the area. Did the hoax actually take place within Area 51?

I'd say that it's just a chance. However, it is not a fact. The Area 51 military base is extremely hidden military base, and the authorities cannot take the risk to test any speculative assumptions. If there is no physical proof the question cannot be resolved.

5.The astronauts took crystal clear photos of the moon's surface, such as in images released by NASA. The company manufacturing the camera said it was difficult to get pictures with such clarity since cameras were mounted to the chest. The user had to shift his body to the direction the direction he desired to snap the picture. In the spherical atmosphere of moon which has a gravity that is just one-sixth that of Earth the body must be moved in such a way to get a sharp picture is almost difficult. So, in these conditions what is the method they used to take thousands of pictures in a single shot ?

It is my opinion that practice will to make a man better. Are you sure you could have walked to the moon, without practicing the technique? It's impossible to tell which 1/6th of Earth's gravity is favorable or not. Perhaps they're finding it more easy to operate their camera with the moon's gravity compared to the Earth.

6.The temperatures range between 2500C during the day and -2500C in night when the moon is in its orbit. The clothing worn by the astronauts can barely endure temperatures of temperatures of 150° Celsius. They returned safe. How did this happen?

It is impossible to know the answer without testing the theory. We first need to know an accurate night and day temperatures of the Moon and develop a thermal simulation on Earth for testing the space suits that they use. The rest is an unanswered question.

7.When there was a sun in front of the spacecraft, the front portion of the spacecraft would be covered by the shadow. It is because there's no atmospheric layer at all on the Moon. In our Earth light sources, the lights scatter due to suspended dust particles that are present within the atmosphere. Because there isn't atmosphere on the moon and the issue regarding suspended dust particles appear to be unsolved. So, there is no air and there is no light scattering taking place. Therefore, light coming emanating from the back of the spacecraft will not be reflected onto the front of spacecraft. It was however observed that the front part of the spacecraft was evident, particularly the more recent USA in the spacecraft.

One of the problems with this notion is that even though it is true that the sun provides the principal source of light for the moon, it's not the sole source of illumination. The other source is the lunar soil reflecting the sunlight's illumination. The lunar surface and the soil feature even reflect the sun's light up to Earth. A similar environment and situation was created to verify the truthfulness of this claim and the results showed that the lunar surface reflected sunlight on astronauts as well as spacecraft.

8. When the astronauts touched down on the moon with the spacecraft and he was there, he was placed in the shadow of the

spacecraft. How could the spacecraft's shadow be evident on the image?

It's the same answer like the previous. Through experimentation the results showed that light is scattered precisely like the photos.

9. There were no signs that radioactivity had a negative effect on the bodies of the astronauts. Radiation from intense radioactivity must be able to penetrate the interior of the spacecraft, which was not properly protected (as as per the researcher). It is likely that the astronaut was very impacted by the radioactive radiation. What is their method of surviving the dangers of space radiation?

As per NASA spokespersons that the spacesuit and rocket have been designed to shield their occupants from harmful radioactive radiation.

10. It can be seen in a few photos that the crosses in the lens of the camera were

covered by moon's surface objects. Then, what exactly does it refer to? What is the way to make it feasible? Could these images be altered?

We don't know the answer to this question.

11. The spacecraft made its landing on the Moon through the blowing of upwards gas. It isn't possible to do this on the lunar surface since there's no pressure on the moon. These kinds of landing tests were unsuccessful even on the Earth's surface. What is the way that a lunar module can land in such a short time on the moon?

I'm not sure. What is the reason that in the absence atmospheric pressure, air resistance or friction, the lunar spacecraft landed on the lunar surface in such a smooth manner? If you know the answer I'm intrigued.

12. There there was no evidence or evidence of "blowing dust" below the spacecraft that it had landed on. This could be a possibility

if the lunar soil Moon is quite difficult to work with. However, the photos clearly display the footprints of astronauts on the lunar surface. It's a contradiction since one could leave a footprint on soil in the case of soil that is hard. The reason why there isn't a blast crater under the moon's lunar module?

Fine grains of dust on the lunar surface get swept off from the lunar lander. They quickly fall back because of the lack of air. So, there isn't an obvious blast crater under the lander, or on the bottom of the spacecraft.

13.Most of those involved in the Apollo mission passed away under bizarre circumstances. Could it be that they died due to knowing the mission too much?

There is no way to be sure. The report of 100 pages written by Gus Grisson was never been discovered after his death.

14.No expression of happiness on the faces of three astronauts was observed after the successful mission.

Yes, I've seen this in a few photos as well as videos. The excitement level following the return of the aircraft seemed a little suspicious. However, such speculations could land one a slap on the forehead. However, I can't deny the fact that I've observed them in a happy and celebration mood as well.

15.Why why were there no stars visible in the moon's sky in any of these images?

Anyone can answer this question. And even on Earth this same concept can be applied. It is impossible to take a picture of stars as well as the Moon by using the same setting that are set in the camera. The exposures of the astronauts in moon are daylight shots. The moon's surface was illuminated by the sun. The astronauts wear shiny white spacesuits that reflect a lot. The camera exposure of the cameras of astronauts were too brief to take pictures of the lunar spacesuits as well as the moon's surface. It was also not enough to capture those stars that are comparatively dim.

Denying America's huge advances in space exploration and the belief in the myths that surround it are "more of an ideological thing--a political thing--than it is a scientific thing," Fienberg says.

There are a variety of arguments which need to be addressed. Bart Sibrel found very highly classified footage from Apollo 11 (don't know how) in which he stated that CIA agents directed the astronauts on when to speak or not speak. According to Sibrel that the CIA even has third and private audio channel. This prompts the crew to react to Mission control after only the four-second time limit has passed, in order to create a illusion of an increase in radio delay. The goal is to appear further from Earth from where they were actually.

A few conspiracy theorists also claim that the footage of the moon landing did not look authentic because Kubrick did the filming, however it is a lie.

In light of these observations is that it's evident that nobody was on the Apollo mission actually landed in the moon's surface. If the Apollo mission actually been realized it would have led to numerous moon landings. Thanks to the advances of 1969 and if a moon landing could be made, the modern science of today and technology could have opened the way to a myriad of moon landings.

We can surely think that the lunar mission was by far the most costly film produced in the USA.

We cannot ignore the truth that we have no knowledge about the subject. The only thing we can do is make our opinions from what has presented to us. The intensity that Van

Allen's radioactive belt. we reacted to it since the evidence was presented to us.

You are free to form your own decision regarding the Moon Landing Conspiracy. My mission is finished by providing you the facts in the proper arrangement so that you can better understand and formulate an informed conclusion.

Chapter 5: Moon Shadows

The very first reason that pops out and I believe the most significant is the idea of

Many shadows are visible in these photos. If you understand shadows and light, you'll be aware that shadows are caused by an object that block the illumination. As an example, if a bright shining light were to shine over you, your shadow could appear to be in front since that's the part which is not affected by the sunlight. The moon should be just one light source, the sun. That means when the sun is in its one location, then all the shadows should be facing in the identical direction. It doesn't appear to have been the situation with NASA's photos. Here are a few of the finest examples of bizarre shadows.

Shadows that are facing the other direction. This is not a problem for NASA

This is a great illustration of how shadows within the NASA moon photos don't have any logic. The shadow of the astronaut's perspective is towards the left, which means the source of light is to the left. However, the flag's shadow is on the left side, that would be in contradiction to this. What is the possibility that we can have two shadows facing in the same direction and only have one light source? We cannot, and it's not feasible unless there's more than one light source This would however contradict the official claim that they were in the moon's orbit because the sun would be the sole source of light which means that all shadows will have the same direction.

The moon landing image that is iconic. The iconic photo is a reminder of how bloopy this whole scene is

It is my opinion that this photo may be one of the most significant for proving the authenticity of a hoax. It could be described as "the smoking gun". The image is a classic which is why and other people, decided to put it on the front cover picture. This is what makes this image so significant. What you're looking at is a photo of Buzz Aldrin, reportedly photographed by Neil Armstrong while they were in orbit. The reason why this image is causing such controversy is because of an obvious flaw it is linked to light sources and shadows. If you take a closer look, you will see that the image of Neil making the picture He is in the in the middle in the visor. On the left, and further into the visor, there is Buzz's shadow. Buzz spreading to the left and right of him. There's a single issue that is unavoidable in this picture. If Buzz's shadow Buzz is visible in front of him, it could mean the source of light was from behind. If you've a good understanding of light and the way our eyes work, then you be aware that in order to view the reflection of light, it would require to reflect off of the object. If only sole source of light is behind him, making the shadow appear to be in front, what is the way to discern the reflection of the object in front of the man? If there was only one source of light, it means that the face of the astronaut is dark, since his back block the light and creating a shadow that is visible in front of his face. It's not logical as it's not feasible ......unless it is true that there were multiple sources of light. That is in direct contradiction with the fact of being on the moon, and that the sole source of light was the sun. It is impossible to have light emanating through the back of an astronaut and simultaneously seeing the reflection of the visor of his head, which will have a slant towards the illumination. This seems to me like a simple idea and logical, except for the fact that I've misunderstood the notion of shadows and light. I am left wondering the level of ignorance people had in the past 40 years that they could not be bothered to consider the obviousness of this error. Sometimes, ignorance can be bliss.

What is the possibility that we could observe Buzz Aldrins' shadow in-front of him and also seeing the reflection of light back into his visor that should be at night? Perhaps Neil was using the flash. ......

How come we see the astronaut's legs even though they're hidden in pitch black shadows since the sole source of light has been shut out?

The image may not appear extraordinary, but to me, it's among the finest examples of why things don't seem to make sense. Take note of the fact that the top part of the spacecraft is visible however his lower part is covered by a huge shadow. Be aware that there's suppose to be just one source of light which means if anything interferes with that light, the shadow is dark, because there's no source of light that could illuminate it, as only one source has been shut off. What is the reason that the legs can be seen? This isn't like an earth shadow, which would result in the legs being covered in shade while other sources of light have illuminated them. Be aware that there is only one light source, there must be no illumination whatsoever for the astronaut's lower half.

This to me proves a single that there were more than one source of light shining on the astronauts. It would be impossible in the event that only the sun were visible.

Therefore, it would appear to suggest that the photos and video were not made by moonlight. It would be to be more appropriate that the shadows are the result of Hollywood kind of lighting, suggesting that the photos were actually taken in an studio. For people who are adamant about the state it will appear off, but there's additional evidence to point toward this idea.

The Flag...

Iconic Image of Buzz Aldrin seated in front of the flag, which seems to vibrate with the breeze, but what if you knew that there's no moon breeze?

Another key element that those who believe in conspiracy theories point to as proof of it is a hoax is the flag placed in the name of Neil Armstrong and Buzz Aldrin. One major point to be debated in this case is that the flag appears to move when it is tossed in the wind. One minor problem is the idea that .....there isn't an moon's atmosphere, and .....there there isn't any wind. What is the reason this be the case that the flag appeared to demonstrate the effect of the wind? Many believe this hints at the possibility that there could be a natural source of wind where the flag was filmed the "monumental moment in Earth history" or artificial sources of wind, like the kind of fan. I'd like to take an objective look at the situation and offer my own hypothesis that I think could explain the motion of the flag. A crucial point to remember is that as I stated previously, there isn't an atmospheric atmosphere in the lunar surface. Given our current winds and gravity, it's almost impossible for us to even have any idea what would happen without an atmosphere. However, I have heard an idea that caused me to wonder if the astronauts actually did reach moon and experience the effects of no gravity and no atmosphere as they erected the flag, they would certainly have moved it after they put it on the earth. Without an atmosphere, even the smallest move could trigger waves or motions that may last forever. If this were true, other evidence seems to outweigh this potential proof.

The absence of stars...

Take a look at the number of stars can you see in the sky when you are looking up at clear nights. Does it not seem unusual that the sky is to be completely devoid of light despite the fact that astronauts are much closer in the sky to stars than the ones we see on Earth?

This is the next an obvious and most convincing evidences that suggest the moon landing could be an illusion. Take a gaze up at the night sky the amount of points of light that can be seen. The moon is a huge source of light. In each section of sky, there can be hundreds, or perhaps thousands of stars to be seen. So, why is it that in NASA images, the background appears totally black? There doesn't seem to be any visible light on the sky. This is something I am unable to comprehend since surely, Moonlight is a lot closer to stars. I'd suppose that the stars' lights would be much closer to the moon than they would on Earth. But with NASA photographs, that doesn't appear to be true. It is difficult to provide any scientific explanation that explains the absence of stars. I'm sure there is someone who is able to provide one. I'm only able to suggest an idea that the majority of individuals are now assuming it is devoid of stars since they're not located on the moon. What you see behind you is type of vast background, black or curtain. It is my belief that the reason behind this is that, of all possible things in order to reproduce in order to make the appearance real, I believe that the stars would be the toughest to duplicate.

It is virtually impossible to duplicate the appearance of stars that are far away, and more difficult to put them together to give a precise depiction of where the stars are supposed to be. However, any future developments and discoveries in telescopes could confirm their claims. Like the other sources of evidence, this could be another indication that the footage wasn't captured on the moon however, it was shot in the present on Earth within a studio. I have a personal hypothesis about where the filming took place and I'll leave the details for when I have more time.

Chapter 6: Moon Dust Is Not Present On The Landing Module's Feet

This is a bit small, yet it is an excellent point. In the photos, there's an additional strange feature; it appears that there is no dirt or dust on the feet of the landing module. Although this may not sound important however let me clarify. If the astronauts actually did actual fact reach the moon, when they returned to earth, they'd need to utilize some kind of booster, like a strong acceleration from engines to slow down their speed? This would let them get safely to land instead of being smashed into fragments. However, there is one evident error that was committed. Recall that, on the moon, there is no gravity, and there is there is no atmosphere. In photographs we've seen, the moon appears be surrounded by dust stones and other pieces of debris. Based on the rules of physics and gravity, when the blast struck the moon's surface it will have caused the debris to rise up into the air. This due to the gravity laws, would be thrown down just as fast in the air. But this wasn't on Earth There is no gravity. Imagine this blast hurling all the debris yet again, only this isn't gravity, and there is no wind nor atmospheric pressure. The debris will not drop straight down again but instead, I think it could float around due to the lack of gravity. In the end, I think that the particles would fall on the ground, which is the point where the issue arises: why is it that after blasting this debris into the air that it didn't come to settle back at the module's feet? This to me isn't likely. In trying to make the appear authentic They had not considered several important elements. It is possible that when they decided on this route to take, they didn't think that 40 years from now Joe the public would be able to examine these images to find any errors. I believe this is an example of a mistake.

Note the super sparkling feet of the landing module. Amazing considering that it was blasted into a slick area and did not manage to collect a single speck of dust.

Nixon's "call to the moon." "...

The president Nixon was the first president to make the longest distance call of his life. Amazing when you consider how we today lose signals and struggle to connect with to the other side of the nation, and yet in

1969, Nixon was able to make an international call to a place which was 238,855 miles from our shores. Take a look at his adorable look! ......the the lying prankster!

One last thing I'm going to make is one which puzzles me. Although I do not consider this to be an absolute evidence it was a fake It certainly makes it difficult to find answers. It's that the president Nixon was alleged to have called to astronauts, and then spoke to them when they were standing at the top of the moon. There is a major issue in this regard. How do you explain the year 1979 that Nixon could make an international call that traveled thousands? of miles, and however, in 2013, despite the latest technological developments, we still have a problem with making calls to different regions of the same nation? It seems like there are two possible explanations one of which is the fact that Nixon was the one who actually placed the call. This could suggest that the government has and always have had advanced technology than we're in the know. Imagine if they had the ability to call the moon back in 1969, what technology do be in their arsenal today? Another reason is that Nixon didn't make the phone call to the moon however, rather to a Secret place in the middle of Earth. It is impossible to say with certainty the truth of either explanation however I am convinced that regardless of which is correct Both have serious implications.

What's the reason cameras don't freeze...

I'd like to mention another aspect, but I didn't think it worth making one of my major issues, yet I thought it was worthy of mention. The reason for this is the type of film employed to make the cameras. Photography and cameras in 1967 weren't the level of sophistication they have nowadays, and they were incredibly basic and used basic film. The camera and the film was specially modified however, this raises the issue of what made the film and camera capable of enduring the extreme temperatures? Keep in mind that we're not talking about the cold of Siberia or even the Antarctic in any way, but rather, it's a completely different world. The kind of frigidity we are talking about is something that we cannot even imagine even a fraction of a second. you can only look at it in comparison to our current knowledge of temperatures in cold weather, that, as I mentioned earlier it isn't even in the same category. It is important to remember that astronauts didn't only have to wear suits to safeguard their bodies from the deficiency of oxygen. They also needed for them to live in the extreme cold. Not just frigid temperatures that could be a challenge when on the moon. Even when temperatures drop to the minus of 153 degrees, they also can reach the temperature of 107°C, so that the film had to be able to endure all temperatures extremes, which I'm sure would difficult to create even in 1969. Take a look at the modern cameras and if you had to bring them into a location where they were exposed to some of the severe weather the planet can give, including the hot desert or coldest mountains Do you think they could withstand the elements? Like Nixon's phone conversation, this makes me to believe that the call was fake, or the government owned the film of some sort that was far more sophisticated that what we is aware of.

Photo editing evidence...

One last item to note is two NASA photos that be clear proof that there was editing. One of them shows the rover as well as other objects. The cross-hairs which are apparent in the other pictures are hidden behind a portion of the vehicle. It's quite odd because the cross-hairs are embedded in the lens of the camera, so they must appear the front or on top of whatever is being photographed. However, as always

NASA photographs do not appear to be in compliance with any logic laws.

The cross-hair of the camera can easily be observed disappearing behind the rover's part of the rover. But how do they achieve this when they have been etched in the lens?

Chapter 7: Close-Up Of The Cross-Hair That Is Disappearing

Yet another illustration of NASA's mysterious disappearing cross-hairs

The second image is like the previous one and depicts a different object, namely an electronic radio with a dissolving cross-hair. This is a must, as the images I have shown are crucial and there is none way for these crossed-hairs to appear on the image's top can appear to disappear in front of an object. There is a method of .....photo processing. For me, the primary reason is the image you see is a photo that was photo-shopped by placing it over the original photograph. That is the reason why the cross-hairs which ought to be at the top of the photo have been spotted slipping under the subject matter. It is important to ask: What is the reason they would need to create fake images? If they did really go to the moon and all is just as it appears, why would they need to alter images? If they really never visited the moon, and it was simply a fake in an elaborate scheme to snare huge amounts of money put into the mission. For me, this is just further evidence that something's wrong in the moon's official story and the images. For those who believe in what the government says, I'm certain they'll come up with many arguments for the reason for this, however to those with even a half mind and a basic knowledge of photo editing, it's pretty clear that there is something wrong with the photos.

The use of props...

The next point is props that could be used from NASA. The photo below was one I'd been able to see numerous times however I had not yet understood what the significance of this image. When I started researching the moon hoax plot that I decided to look at this photo and what I found inspired me to laugh but then be angry about how fat the people are really.

What was the way that a perfect "C" find it's way onto the moon stone? Perhaps an astronaut took part in some art work on the moon?

The image you see in this photo is moon rocks. On the surface, this may look plausible, but if you don't examine it very closely, then you'd think it was an enormous moon rock however, why is there what appears to be a huge "C" on the surface of this rock? Perhaps the image you're seeing might not be an actual moon rock but actually a huge paper-mache rock? It could be a prop? If it's really created by a human, surely NASA would not be foolish enough to put a clear word on the surface of the object? In fact, if you take a look at other evidences and in the event that NASA wasn't so stupid there wouldn't have been this debate. It is still unclear if it is indeed fake, what else could be authentic? It could be possible that they merely sprinkled a portion of the desert in water as well as tiny stones and rocks to give the illusion of the moon's surface? It could be that they did not want to utilize large earth rocks in the event that they were noticed by an expert, so they made a few pieces of paper mache, labeled by letters, so that they could recognize where they were and that some stagehand did not remember to wash off the letter 'C' or conceal it? If it really was an prop, and I think it is, may they have made use of alternative props? Did any moon footage taken actually from the moon?

Simple mistakes...

In the meantime, I'd looked at so many photos and videos describing how this might have been a hoax that I was already decided that the whole thing was a ruse. It convinced myself. The video featured two features I'd not had before, which as you'll realize is unusual if you've been to any of the many tens thousand moon landing hoax movies available on the web. Here is a link to these videos, as and include a few pictures, but let me to explain briefly what fascinated me about the video. The video featured a portion that showed astronauts aboard the rover going to the west during their initial day. They come to the top of a hill. The video is paused when he reaches this point. He mentions a particular rock he believes appears to be a mouth of a crocodile. In the following video, the astronauts take another vehicle, but the time it is revealed that they're heading to the towards the east, to investigate. When the user pauses the film, and that same mountain appears instantly identifiable due to its "crock rock". Additionally, this time the moon buggy appears on the top of the hill, while it was not present at the beginning. What is the reason in two days moving in opposite directions, they end up at the same place? It's due to bollocks. If it weren't, they wouldn't have to spread the truth and attempt to fool us. But basic observation and level of intelligence can tell that it isn't the case. I've considered this many times before. How could they have such a bad rap? They seem to have misled many individuals, but it's not widely known and it appears to have covered lots of the bases, however often things are overlooked. It happens to serial killers. Regardless of how clever they may be and no matter how diligent they are, they will always fall into making a mistakes and falling short because there's no way to account the way other minds think. No one has the capacity to be able to think of anything. Another important aspect of this film was a set of photos where it became clear to me astronauts were in an appropriate harness that could lift them to the sky, giving the illusion of no gravity. The most effective example occurs when the cameraman spots an astronaut on the background and it appears to be falling. When the other astronaut reach for help, it appears that the first one corrects himself in an impossible manner. On closer examination, you will notice that the astronauts don't ever touch, and it appears that some force, which is not visible is pulling the astronaut back on his feet. The force is invisible is in this scenario is most likely wires and an e-harness. There were examples similar to these that helped me to make the decision on what exactly happened.

This photo was snapped in the first day of exploring to the west. The astronaut is next to it. highest point of the landing platform can be seen. Take note of the rock that is in the hill, which resembles the mouth of an open crocodile

This photo was taken during the second day of exploring, but this time in the west. You can see that the exact identical 'croc rock' remains apparent, but this time, there are two astronauts on the ground and the landing module has disappeared. What is the possibility that in two days with two different directions, they'd finish up on exactly the same hill, with exactly the identical rock? This isn't likely ....its an untruth!

Chapter 8: Conspiracy True?

To Teresa

We all have a personal obsession in the Moon.

Education and Patriotism

We Americans learn in schools, "The United States went to the Moon a long time ago, but it was so barren, totally nothing to see on the Moon, and way too expensive to go back." I've heard that argument previously. It's like we have been taught to believe"that.

Why could you explain that the Moon during the past few decades, be the sole field of study that we have stopped exploring? Our explorations slowed down of the Moon as well as our reasons included:

1. It's too risky not to go to the back.

2. There's no real reason to go back; it's the exact as on the surface of the Moon.

3. It's way overpriced.

If those are the main reasons that no other nation has taken a trip to the Moon for more than fifty years, I'd be pleased to share the following reasons:

1. If this is as dangerous How could anyone ever pass away while flying towards the Moon? It's the death of those who were on the flight practice. This doesn't mean that flying there is risky.

Actually, it was not at all scary! It was a golf course in the Moon!

2. Is it boring? A bunch of Moon rocks? There's nothing to look at? Is that really true?

It was quite a bit of trouble and finally landed in a small area where you played golf. You did not do any scientific research since you were not researchers, you were just militaristic patriots and decided it was unnecessary to return?

There's no need to return, as we've seen all it can offer.

3. Too expensive? Really? billionaires are spending what it costs to go into the low-earth orbit but it's a lot of money to duplicate the feats NASA used to do with the 1960's technology for a lander made of aluminum? There's no reason to believe it would be too difficult to obtain the chance to take a golf course in the Moon. Which billionaire could not have had the chance to play golf on the Moon in the past?

If you're listening to me perhaps they didn't go due to it being costly and risky. If I'm correct, this could be one of the most effective cover-ups of the past. Some of the top researchers in America appear to be taking part in the cover-up.

If it was a major fake, it would suggest that the president Nixon was the one responsible for greater than Watergate. In the 1980s, the U.S. had enough motive to establish its global supremacy throughout the Cold War -- even the greatest propaganda film. How better to reinforce its position as a leader then to launch some of the strongest rockets on earth to space. These rockets were, as you may recall was always live broadcast throughout the 1980's.

I vividly remember NASA launches in my youth as they were often able to preempt regular TV programming. What was the truth? The reasons behind the Moon landings might be a bit defensible however I am still blown amazed by people who are reasonable and willing to discuss the issue. Every time I think about it, I think, "Eh, what does it matter anyway?" My personal experience has convinced me that it really matters. Truth matters.

This isn't to say that the rockets are fakes. These were authentic rockets. They had millions of witnesses at every launch. It is not my opinion that they were able to fake the launch of massive rockets. This is a red herring.

Flag-waving in the Moon controversy is nothing but a false premise. The shadow lines can be a red herring. The Moon doesn't have stars in its sky--a red herring. They've all been widely disproved for many years at least to a certain extent, which means there's plenty that the American public can view to keep them from being fooled.

There are numerous kinds of misinformation available. "People who believe it was faked say the flag was waving. It wasn't waving; it was bouncing on an aluminum rod. Therefore, your entire argument is disregarded."

There is no credible Moon space landing conspiratorial theorist (what an amazing paradox) will employ the flag's waving for an argument. It's a bit absurd to base your whole conspiracy idea on the "flag waving" that probably wasn't even moving.

There's plenty of evidence that is legitimate online on YouTube nowadays, along with some obviously absurd arguments that are simple to ignore. Take the Flat-Earth conspiracy for instance. It wouldn't surprise me to find out that NASA was actively combining moon-landing videos together with flat-earth video clips on YouTube. It is possible that they have cooked up the idea of a flat earth. If you're able to cut through the maze of misinformation on the internet I guarantee an exciting adventure in the realm of conspiracy theories. The fake news was out there. I did take the blue pill.

First Attempts

The third version of the book. I wrote under my pen name Teddy Freeman for the first two. I was convinced that I was onto something so potentially-sensitive that I might face censorship. I'm aware that it's true that the United States has seemingly always been the defender of free speech. And if I was to be right, this could be a matter of consequence.

As a teacher who frequently teaches American historical events I am conscious of our country's capability to deceive and the requirement to maintain control over conversations when needed. In extreme instances I'm sure the need to appeal to scientists' feeling of patriotic duty.

If we never went to the Moon It is a disappointment to all humanity. Political consequences could be grave, the scientific consequences more. A widespread acceptance of this reality could one day cause us to doubt everything we've ever witnessed on television. This could also cause the public to doubt each celebrity-level scientist in the near future. My admiration towards Bill Nye and Neil deGrasse Tyson has been affected, not simply due to their belief that they were the first to believe that the Moon landing is real, however their logic seems to be a bit simple.

As per Moon landing people, the video could not be faked. They claim this is proof that the Apollo missions as genuine. A keen ability to evaluate the evidence which is in opposition to conventional wisdom is an oddity when prominent television scientists are asked questions about their experiences with the Apollo missions. It's what you get after you have spent years fighting each and every aspect in "amazing science" and very few hours analyzing critical issues.

They aren't the best people to talk to regardless. According to me, they are as credible regarding this issue like the folks who appear on Mythbusters.

I've been asked many times "What does it matter anyway?" Take a moment to consider the importance of that issue. Science will be affected over the next few years.

Scientists think that scientists believe that the Moon was once home to an electric field, which is because of the location of the elements found in Moon rock samples returned by Apollo. It's not much simpler to understand once we know this? Moon rocks that were collected through NASA can be described as Earth rocks?

"In the 70s Scientists subjected recently arrived specimens from Moon rocks to a flurry of tests. They were surprised when they found that a few rocks had magnetized. After the magma had cooled, and solidified in these stones, they was exposed to magnetic fields. This was a bit odd because the scientists were aware at the time it was believed that the Moon was not surrounded by a magnetic field as strong as those that surround Earth at the time, or even. However, the rock samples suggested otherwise in a number of studies. these samples have since proved that there was a magnetic field on the moon. Moon certainly had an electric field that was present billions of years ago. It could be similar to Earth's current one."

(Marina Koven August 9, 2017)

Here is the link to the Bart Sibrel radio interview. If you're really keen, like I am in finding out the facts about the Apollo Missions I would suggest listening to the whole interview.

Chapter 9: Occam's Razor

The principle was formulated in the name of William of Ockham (1287-1347) who was a friar as well as a scholastic philosopher and theologian. The principle is that if there are more than two hypotheses competing it is the one that rests upon the smallest number of assumptions must be selected.

A Pointless Conspiracy?

I was of the opinion that being a part of the U.S. government staging the Moon landing seemed like a bizarre passion that is just the thing for bored people. To a degree, it does. What person would want to watch hours of NASA footage? Yes, I did and then I surprised myself when I came to the conclusion that United States did, in the end, fake it.

For me, the most decisive factor was the motivation. It was vital to America U.S. to be successful in this undertaking. There was no way to risk an ominous Moon accident. Like all travel in space there was the possibility of loss of human lives was likely. Strangely, stepping through the Moon appeared to be a simple task for the largest nation that is located on Earth. It is not necessary long to realize that countries with power can be powerful in their propaganda.

The United States was desperate to restore its international power following the humiliation of Sputnik. Actually, the Russians had been proving in the United States in all sorts of ways. They were the first satellite made by humans, the first astronaut to orbit, very first man on the moon and the first woman to be in space, the very first dog to orbit... It's a long list that is endless. Should you were to imagine that the United States tried something as bold as attempting to go to the Moon and did not succeed it would have repercussions that were huge.

"It is possible to go to the Moon," researchers of the day were unanimous in their belief. A minor snag could jeopardize security for the nation's most powerful nation on earth. A large part of our United States' advantage since World War II has been technological advancements.

Nixon

Nixon was the president during the time. Nixon was the president at the time. his involvement in the Apollo Missions. Most likely, he was among those who came up with an entire concept of creating the Apollo Missions. If anyone in the past president's administration was adept at this

the one who did it is Richard Nixon.

Clinton

In the 156th page of the autobiography of his, My Life, former president Bill Clinton hints that even his own questions whether it might be played out. The language is shrewdly concealed by the small breadcrumbs he left behind

"Just one month ago, Apollo 11 astronauts Buzz Aldrin and Neil Armstrong had left their partner, Michael Collins, aboard the spacecraft Columbia and had walked across the moon. Moon...The old carpenter questioned me whether I believed that it was true. I replied that I watched it on TV. He was not convinced; he stated that he did not believe, even at all, and the television's 'fellows' made things appear real, but they weren't. At the time, I believed the man was cranky. When I lived living in Washington I was able to see certain things on television which made me wonder if the man was years ahead of his time." William Clinton, J. Clinton. My Life

Van Allen

The Van Allen radiation belts are two regions of charged particles. They present a serious obstruction to any future space travel. they're situated between 9300-12400 miles over the Earth's surface. NASA acknowledges it is extremely risky for any human being to walk in that region.

It's not surprising that this did not happen during 1969 or any other Apollo mission. The fact is that all the Apollo astronauts lead long, well-nourished lives, but they did not suffer from the effects which are typical of radiation poisoning. It is possible that we didn't have the ability to pass through the belts in a safe manner back in 1969, however we had the capability to transmit a virtual simulation.

Houston

Could a prank that this large be difficult to fake? It's not in the least. Technology was so primitive and insignificant, how is this not apparent to everybody today?

A few in the upper echelons knew about the fraud, which included the astronauts. All the rest: researchers, technicians from Houston and broadcasters who were able to accept the live feeds believed that they were real. And so did the majority people in homes and at schools around the globe. Everybody was watching the exact identical "live feed."

Is it propaganda or science?

It is impossible to duplicate it?

A Distant Playground

Everybody has seen the footage of men playing across the Moon. They were playing golf. They played on their go-cart, which appears to be still in the park. They would do jumping jacks as well as all kinds of other antics. I'm sorry, but isn't the Moon not a tad chilly, this would mean that the tiniest tear on their clothing could result in immediate death, surely? It's never an issue during the fun stumbling and hopping about in the Moon.

The video evidence online is astounding. The video clips that are analyzed within the hyperlinks I provide are authentic NASA videos. I'm not going to get into too much specifics because the video really speak for themselves.

If you're searching on the internet, keep at heart that Moon landing video conspiracy theories are typically associated with other, typically bizarre conspiratorial theories. It is important to be able to identify those who promote the "flat-earth theory" disinformation--it is all over YouTube. There is truth to it However, it is important to remain solid in your the logic. The Earth isn't flat.

Sibrel

I enjoy Bart Sibrel's tenacity. It can be frustrating and confounding, however I'm not convinced he's insane. I believe he's correct. There are clips from his film A Funny Thing Happened on the Road to the Moon on his site, Sibrel.com.

There is no doubt that every government, not just those in Washington, D.C., has the capability of creating an appearance for the people.

Disinformation:

Intentionally false information, often deliberately spread (as as by planting false rumors) to influence public opinion, or to obscure the reality.

Chapter 10: Easily-Overlooked Conveniences

To train astronauts A exact recreation of the serene sea that was the site of landing for Apollo missions, Apollo mission, was dug from the desert of the American southwest. Wouldn't it be nice?

According to NASA they found an reflector that was located on the Moon's surface. the Moon that is frequently used by scientists. This made it possible to determine the exact distance between Earth to the Moon. Earth towards the Moon. What is the reason why nobody points out the fact that laser beams are able to accomplish this without the need to have a mirror that is special at the other side? Laser beams can measure a variety of near-Earth objects with this technique. Isn't it convenient that the Moon is already an especially-reflective surface.

Easy to Prove, Either Way

There's a simple method to silence all the people who are a conspiracy nut, such as me. Use one scope, one at a time, over the Sea of Tranquility. Zoom to the right. Display the evidence that has been that was left by the footprint. This footprint, according to what we've been told, will stay completely intact even in space. In case you're not aware, the Moon is not surrounded by an atmosphere consequently there's nothing that can act on it.

If we hear that the telescope is not available I would guess that this is what we'd think to hear if it was an attempt to cover up. Astronomers have the ability to focus on particulars far from the scene. Public attention is frequently focused on the notorious place of landing in the Sea of Tranquility.

The past NASA initiatives that travelled close to the moon, having modern cameras, have never believed that it was necessary to be able to zoom in and pay homage to this holy place for human exploration.

NASA spacecrafts have frequently employed the Moon to sling-shot space. There has been no single photo from that historical site. It's a bit odd. It's not what you would expect from an Moon blackout by a government that is worried the people won't be able to trust the information they receive on TV.

I am aware that the Myth-Busters team did a few shows about it and they were looking for an evidence which can easily be disregarded. There are tons of ridiculous arguments that it's an effective way to distract away from the truth that I've already mentioned.

Adam Doesn't Ruin Enough

I am a fan of Adam Ruins Everything. It seems to me that Adam Conover has an open thinking mind and usually is to be only interested in details. I wondered what he thought was logical in his assessment of the Moon landing, when he stated, "Given the film-making and lighting technology at the time, [faking the Moon landing] actually wouldn't have been possible."

Is that not even remotely possible? Do you believe me? How difficult is it to apply common sense to this subject? It's as if everyone is brainwashed we believe in this message. It's very likely. To illustrate, take a look at the Moon scene in 2001: A Space Odyssey. The idea of saying it was impossible in 1969 is a slap to the public's intelligence.

If there's a desire to believe and a population that believes in Richard Nixon, there is the soundstage, as well as grainy film footage that's believed to be "impossible" to fake.

Do you find it difficult to realize that it was not possible to make up it. Moon landing? If you've decided, like myself, that it was possible and then you start to realize the possibility that Adam Conover seems to be in the matrix in this particular instance.

It is only possible to believe it's impossible if will put the entirety of your conventional wisdom aside and go on blind conviction. The analysis of the facts can lead to a different conclusion.

It's highly suggested that you not accept anyone's words for it. Check out the footage, observe the circular logic employed to justify the Apollo mission, and take your own decisions.

Justifiable?

Although it was not the most famous telecast to deceive the world It was equally brilliant as it was defiant. If the United States government did fake the Moon-landing footage from 1969-1972, then it was a genius strategy, albeit an immoral and ethically-reprehensible one.

If the film's propaganda is seen in the context of the overall U.S. space program, that is closely linked to and influenced by the U.S. military, one can argue that 1969 represented an end to the Cold War. in the Cold War.

The Apollo Mission era was the base on which America built its identity. United States fashioned its forward-leaning character in the latter part of the 20th century. I don't want to dive into the controversy that have been aired online regarding NASA continue to steal billions through deceiving the public, however there is plenty of interesting information that can be found on YouTube as well as elsewhere.

I'm a bit skeptical of the content you can find on YouTube like everyone else should be. Contemporary NASA video tricks could be another red herring promoted by the Powers That Be to steer people away from the lies in Apollo Missions. Apollo Missions.

The U.S.S.R. was unable to stop the momentum of the United States in the decades after Armstrong's famous remarks. Then, eventually after that, it was the Cold War ground to a slow halt.

In the aftermath of the collapse of U.S.S.R., the United States could have taken an opportunity to justify the need for the lie however, such an admission could have been seen to be a sign of vulnerability. Do we believe in such events once more?

If the government was to acknowledge such a deceit could we have sunk to all to the twentieth century doubting the information we watch on TV? The conspiracy theories of 9/11 provide an indication that the majority of individuals are already in the same situation. To be clear I am not convinced of any other theories than the widely-accepted facts of that tragic event.

The major difference between 9/11 and Moon landing conspiracy is that there is a distinct motivation. I've been able to hear the motivations provided by 9/11 truthers. They are not convincing. They're just too intricate as well as the death toll was in the hundreds. The idea that it was an intentional incident by someone other then terrorists is a bit of a stretch to me.

The Apollo missions were not the same. It is possible to argue that the faking of the Moon landing was a way of saving lives through stopping the Cold War. It's a valid argument for a large portion of the population.

Chapter 11: The Patriots

I am convinced I believe that Buzz Aldrin, Neil Armstrong along with all the patriotic Americans who were involved, were decent individuals. Their motivations were noble. According to me, they may have had a good motives to believe that the fraud was legitimate. What, in all the decades, hasn't it been thought of by any one to think that it's an appropriate time to speak the truth?

I'm convinced that Armstrong may have been told the truth would be revealed in the end. In the event that it did not Armstrong preferred to remain away from the subject. He was regarded as an "reluctant American hero," Armstrong was kept out of the spotlight for a large portion of his time. Moon conspiracy theorists claim that his conscience could not permit him to keep a lie going.

"To you, we've just completed the beginning. We will leave much to you to be done. Many great concepts remain unexplored and breakthrough possibilities for people who are able to remove one of truth's protection layers."

-Neil Armstrong, 1994

A recluse's life is a strange choice for an explorer determined enough to walk over 238,000 miles and be the first person to step foot through the Moon. Did a person of his personality not have felt the duty to keep answering questions till the end of time? The men who have been on the Moon do not seem to be commonplace.

The descriptions he provided would have been a tremendous help, just like any of the astronauts that took part in other Apollo missions. It is interesting to note that the stories of astronauts all have a clear and unambiguous view that the area was barren space with no lessons to draw lessons from.

Search for the Truth

There is no need to believe me on that. In the case of every conspiracy idea, I recommend to look at all available evidence prior to making your decision. My experience has shown very open-minded. If you're determined to show it the other way around then you need to set your preconceptions aside and observe every piece of evidence given.

If I searched on Google "Proof we did land on the Moon," I thoroughly examined every single piece of evidence available. It was difficult to locate a lot. There is no doubt that there was evidence for example, a satellite photograph of the landing area or an undeniable Moon rock.

When I first look up a result I find it appears that The Weather Channel has some evidence. I had hoped to find an image from a satellite from a similar site. However, this is the story of a film director who says the photos that were seen in 1969 may not be produced. Really? What a silly argument. Utilize your common wisdom as a mature adult. Do they have technological capabilities to make the Moon launch in 1969? Yes they did! One of the most popular films in 1968 is 2001: A Space Odyssey that depicted the moonscape as well as the movement of low gravity with great accuracy.

The NASA Moon footage has high-quality image quality. It's a huge event it would seem that they been equipped with the finest cameras around.

Star Trek: The Star Trek pilot aired in 1965.

The reasons how they were able to produce only blurry footage are ridiculous. You will also have to be prepared for a very unprofessional description of how heavy the cameras were at the time and why that was the most advanced technology of that time.

If the latest cameras were this massive, you ought to leave that lunar rover in the dust. It was used for nothing other being a prop to Moon actions. All of humanity could have been better off with more Moon footage.

The footage was inspiring to millions of people.

Films like this would be scrutinized in detail for decades to come, however NASA is "lost" all film that isn't yet been publicly made available.

Popular Science was the next website in the results of a search. I was hoping to see some proof. However, Popular Science only offered another "analysis" of the footage. The footage wasn't even convincing. Camera tricks like this were common even back in 1969. I'd like to see a smidgen of evidence that is irrefutable.

It must be proof of airtightness that exists somewhere! This is because it's possible that the United States was doing it to show that we were the first! The astronauts are expected to have returned more than photos and rocks that resemble fragments of meteorites. That's right that there was nothing other than rocks.

Popular Science provides a painfully-unhelpful article which delve deep into the specifics of how the Moon dust that was spewed up by the rover may be only happening in the Moon. This isn't even close to the evidence I'm looking for.

When you look up the second result the website referred to as "Universe Today" hurls insults towards conspiracy theorists, but offers neither evidence nor proof. They offer absolutely no information whatsoever. They reference the fact that thousands thousands of individuals were involved in the Apollo project. Like I've said numerous times, everybody was involved in something they believed was an authentic project. The entire control room believed that the project was true.

The research was believable enough, however it's likely that the spacecraft carried no people and the signal receiving, which was also received the NASA control room, was controlled by a small group of people of an audio stage. The simulations were very similar between the computer simulations and the real "event."

Each search result that I came across is identical. The populace is so desperate for this to be the case. It is not as convincing as it has ever been however, the faithful surpass the skeptical.

It is obvious how unscientific many of the well-known claims are. NASA has awaited for everyone to be in the air swaying in the water, eager to make use of similar footage to show that the moonwalk did take place. NASA maintains that they have walked across the Moon in the Apollo period. What is the reason there's no source of proof other than grainy footage as well as high-definition images of astronauts in poses? There's nothing else to prove it which is just the same as the Bigfoot sighting.

I would like to warn you on trying to locate the proof on the internet. It's painful. NASA has nothing to prove. The video clips released by NASA offer a compelling reasons to be skeptical. It's difficult to fabricate We aren't stupid.

"If you were standing on the street and ask every passerby to explain what the facts are and they would all say that the truth is the only thing that is true and it is true. But, this notion is a misunderstanding. We interpret facts based on the truth and reality.

The problem with beliefs is that they may be completely off. Human history is filled with several instances... Human people need to know facts as they require certainties to navigate through the world. We must not forget the human condition."

Chapter 12: What Are Conspiracy Theories?

There's a myriad of websites that cover the myriad of theories. In this chapter that the fact of this as well as the existence of Social Media such as Facebook and Twitter has made conspiracy theories much more popular and popular more than they ever were before, but not at first.

What is a conspiracy? An act of conspiracy happens the case when a group of people come together and collaborate to carry out something that is illegal, unmoral or devious.

Most crimes committed by criminals are due to plots. Security agencies, such as KGB and the CIA CIA and KGB come up with plans, that are required to be secret, and they are constantly in the process of collaborating.

The majority of companies want to keep the confidentiality of their trade, however they frequently employ persons to sabotage the confidentiality of their colleagues and by so engaging in a conspiracy, which is to participating in a conspiracy.

A majority of people have been in a few minor plots like two kids who conspire to rob the liquor cabinets of their parents. Other conspiracies have been committed by the government that could succeed or not. There are two conspiracy theories which have been extremely successful.

The site that was the site of D-Day landings was long before June 6th 1944 on June 6, 1944, the Allied Supreme Command had decided to attack France within Normandy and had devoted their entire preparations to that goal. They had a strong desire to ensure to ensure that the enemy Nazis could not be aware.

The Nazis believed they would launch the war within the Pas de Calais region of France, which was a lot more close to Britain.

While the war was certainly morally acceptable, it was a not in the public eye. The Germans didn't realize they had erred until they realized it was too late.

The growing rapprochement of China as well as the USA for a while prior to President Nixon's spectacular visit in China on February 22, 1972 were held secret talks between the officials from these two powerful nations.

The plan for the trip which has consequences that are being felt is a hidden. There was a conspiracy however, even if it wasn't because of moral motives.

Certain conspiracies can go wrong but.

Gulf of Tonkin Incident: The Gulf of Tonkin Incident in 1964, involving US Navy vessels and North Vietnamese was a strange incident.

Johnson of the USA. Johnson from the USA utilized the occasion to justify an enormous increase in American participation during the Vietnam War. The planning for this participation was in place prior to the war but kept under wraps.

It was a plot, that resulted in only one loss that the USA was able to defeat in conflict. This was a catastrophe.

Watergate The Watergate conspiracy is thought of as the most significant political scandal on earth.

It was essentially the involvement of The president Richard Nixon and some of his closest aides to obtain subordinates to gain access to the Democratic Party's national committee's headquarters within the Watergate complex located in Washington in the year 1972.

The culprits were arrested and, over the following two years, the scope of the scheme was discovered. The discovery resulted in the resignation of the president Nixon in 1974.

Kathleen Kane: In 2016 the Attorney General of Pennsylvania Kathleen Kane was convicted on several charges such as perjury and criminal conspiracy.

Leaked documents to the grand jury to undermine a political opponent. The woman was prosecuted and then was convicted of lying about her actions, but her demise was determined by a jury, thus the end of a remarkable career.

What are Conspiracy Theories? There are a variety of definitions for the term "conspiracy theory.

The most reliable source I've found is an online encyclopedia called Wikipedia. It's a good idea to study the definitions.

The conspiracy theory is the notion that everything shouldn't be taken the face value. This is due to the fact that there are powerful people or entities that control things for their own purposes like financial gain or the power.

Conspiracy theories have existed for as long as there has been. One of the best examples is the notion that were promoted by the Nazis in the belief that Jews contributed to Germany's loss during World War 1 and what they believed to be the Nazis believed was the demise and devastation of the Aryan race.

Another one was the idea propagated by the Communists that any wrongdoing within society was attributed to the monopoly capitalists that had control of everything and mastered the oppression of poverty, oppression and war to earn the profits.

A different, more widely held belief is that there exists an angel of the fallen called Satan who is an unseen supreme ruler of the Earth and everyone who lives on it. He is the one responsible for every evil.

The belief in this is the foundation of many religions.

What Are Some Widely Believed Conspiracy Theories? The death of US president JF Kennedy is thought to be the result by a single gunman Lee Harvey Oswald.

The explanations are viewed as a cover-up for those who view this as a outcome of the manipulative actions of the Cubans and the Mafia the Teamsters and Vice Johnson, Vice President Johnson as well as the Illuminati.

Princess Diana who was divorced from Prince Charles was killed during a collision together with her boyfriend Dodi Fayed inside a vehicle that was driven at a high speed by Henri Paul, a security official from Fayed.

Since Diana's death, it is blamed on a conspiracy with The British Secret Service MI6, the Royal Family, and the Illuminati.

The 11th September, 2001, planes carrying passengers taken by Arab terrorists affiliated with Al Quaeda crashed into the Twin Towers in New York.

The result of this as a result, President Bush was able to justify an invasion of Iraq that caused a lot of trouble all over the world.

This incident has led to numerous conspiracy theories, the majority of that claim official cover-ups. This time, the Illuminati are among the possible conspiracy theories.

Who Or What Are The Illuminati? The Illuminati did exist as a secret society formed in Bavaria around the turn of the century. They were deported at that period and were later blamed as the cause of French Revolution.

They are now believed to be at the root of most conflicts, revolutions, atrocities and even catastrophes.

They've transformed, within the thoughts of their adherents to a shady group of people who want to establish the New World Order.

The group is thought to be comprised of the Kennedys and the Onassis family and the Rothschilds as well as Angela Jolie and Lady Gaga.

A few of those who believe in this conspiracy say that participants must make blood sacrifices.

There is a belief among some supporters that celebrities such like Michael Jackson, John Lennon and Elvis were murdered in the hands of the Illuminati.

The Illuminati are said to have many symbols that include items as the number 666, a double lightning bolt, which was also the SS Nazi symbol, the pentagram and crossbones and skull.

Another emblem from the Illuminati is said to represent an all-seeing eye that is visible of the pyramid, which is depicted in the US dollar.

Why Do Conspiracy Theories Spread So Quickly? These theories are increasingly popular these days. If you walk into a bookstore, you'll see publications and books that question and challenge the official versions of a variety of things.

Every search on the Internet will result in thousands, if not many similar criticisms of the accepted wisdom.

In 2016, Americans have elected a new president, who appears to be a believer in these theories. The most common conspiracy theories come in times of fear and are able to see that people especially those who are in charge are open about their concerns.

The research is conclusive that shows that while conspiracy theories are believed to be true isn't as prevalent today as it was a half century ago, the pace of the emergence of conspiracy theories are much faster. It is because of the advancement of technology.

There are million of people around all over the world who use their phones or tablets to upload photos and videos of their things online. In the sense that they are broadcasting with news.

It is likely that there are many more people who are sharing opinions and perspectives about this information.

The confluence of two things that were never seen prior to the time of humanity it, have led to the emergence of conspiracy theories prevalent today.

What are the causes of Conspiracy Theories? The issue of conspiracy theories is the focus of numerous scientific research studies.

A lot of them come to the conclusion that there exists in the human mind a need to find patterns.

A number of the top mathematical and scientific achievements could be traced to this.

Artificial intelligence programmers write code to do exactly what they want to do. They use perceptrons that look for patterns.

The search for them is supported by reward and punishments, to ensure that the application of the rules is comparable as the way a dog is trained. dog.

When people spot a pattern and recognize a pattern, the following step would be to assume that the pattern could be created by an agent that created the pattern.

This behaviour was very valuable throughout evolution, and even at the beginning of humanity, when mankind was still in the hunter-gatherer stage.

The sound recorded by early humans could just be the sound of breeze or the saber-tooth Tiger.

The most secure course of action was to assume that the saber tooth tiger is the one to blame.

The sound was the pattern and the saber tooth Tiger is the actor in this particular instance.

As a programmer who successfully codes the artificial intelligence program, our response was encoded in the DNA of our body.

We are seeing our operatives in the form of the state as well as the Illuminati as well as the gods of religion, and many other entities when we've observed patterns.

Today, we have every reason to be skeptical of the government's actions large companies, bureaucracies and academia, etc.

It's quite normal to think of one as an agent, rather than believe the reason they offer.

Does any harm come out of Conspiracy Theories? The majority of conspiracies are harmless and nonsense.

If one believes that the Illuminati are the ones responsible for the problems of the world, to create a World Government then such a assumption is most likely untrue.

However it's not always the case that conspiracy theories are harmless insanity.

It is believed that vaccines trigger autism.

The theory suggests that children are injected with poisons that can result in them developing autism.

The reason why the vaccines are being used is to earn money from drug manufacturers, that manufacture the vaccine. They are influencing the federal government to force vaccines onto a naive public.

Because from this conviction, a lot of parents have prevented their children from having their children vaccinated. Research has been conducted to prove that autism does not result through vaccinations.

But, if children are not protected against measles or one of the other illnesses they could be protected against, they might be unable to live.

Nobody wants this to happen to their child or son. It is a risk that removing vaccination on them places the children at greater risk as compared to if they had been vaccination-free.

This is an obvious case that believing in a conspiracy can be harmful.

Is The Moon Landing The Subject Of Conspiracy Theories? If you believe in the Moon landings are real, they are the pinnacle of human achievements.

Amazing organization, amazing technology, incredible courage in fact, the Moon landings symbolize everything that is good and admirable that humanity has to offer.

However, some people don't believe that landings occurred. They say it was unattainable in 1969 to an astronaut to the moon. Moon.

They're certain NASA has fabricated the landings in the form of fake films claiming to show things that didn't happen.

They think that NASA was doing this on the instructions by the federal government, who was trying to show all the world that America has superior technology over Russia.

There is a certain fact that there was a concealing of the truth for the last half century. The Moon landing has become the center to conspiracy theories.

It is amazing how the Illuminati are being dragged into this, but many have a simpler explanation.

The main reason why the federal government might want to pursue this strategy is due to that it is the Cold War. The following chapter will examine this remarkable event in the history of mankind.

Chapter 13: The Cold War

"I lived through that period of the Cold War when everything seemed extremely fragile. In the years that followed, through the collapse of the Berlin Wall, I had vivid nightmares about nuclear disasters."

--Justin Cronin

The Germans started a well-planned attack on Poland on September 1st, 1939. Thus began World War II.

Russia also known as the USSR in the past was also invading Poland on September 17 1939. It then captured the eastern portion of the country that was devastated.

A country which has been victim of numerous invaders throughout the years.

The Soviets were able to invade Poland in response to the Ribbentrop - Molotov Pact signed in the month of August 1939. This pact Nazi Germany and Soviet Russia reached an agreement on a mutual attack.

Actually, Russia invaded Poland as they were in need of as much space between themselves and the Germans that they can.

They anticipated that Nazis were going to eventually attack them.

In the same way, they took over the smaller countries of Lithuania, Latvia and Estonia that are adjacent to them along the Baltic Sea in early 1940 They also started an attack against Finland during 1940.

On the 22nd June 1941, having conquered the majority of European nations, the Germans finally began a major attack against the USSR.

In the span of nearly six months, for nearly six months, the German Army or Wehrmacht were unable to stop. The Russians were a victim of millions of deaths.

But, the Russians defeated the Germans at the end of December 1941 during the Battle of Moscow.

That the same month Hitler who was the German dictator declared war against the USA. It was to be an extremely fatal mistake to Hitler's Nazi regime.

The Nazis revived their war to defeat Russia in 1942. Russians in 1942, resulting in stunning wins. The Russians were retreating in full force.

But they fought back and inflicted a devastating loss to the Wehrmacht (German Army) during the Battle of Stalingrad.

By the time 1942 came to an end at the end of 1942, the Nazis were retreated, and both in Russia as well as North Africa.

The Germans were stricken by another serious defeat at the battle of Kursk in mid-1943, and not to be again in the front along the Eastern Front against the Russians.

They could only go back and accept loss after loss in the same way they had done. The war lasted for another two years, or so, war continued until Hitler killed himself and the Germans gave up their arms in May 1945.

In the course of their retaking of the Nazis during the war, the USSR took over nations that were part of Eastern Europe. The Russians had to force these countries to become communists and then the Iron Curtain fell over Europe.

The Cold War between the communists and the West started.

It was the Cold War was not a conflict in which the armies of the forces from The Soviet Bloc and the West were at war, but rather a state of conflict, however.

The only shooting wars occurred through proxy wars in the Korean War, in colonial or revolutionary wars. They were usually however, not all of them initiated by Communists and against corrupt governments like those of the Batista government in Cuba.

Some events looked like they could have been war-related, like the construction of the Berlin Wall in 1961, the Cuban Missile Crisis of 1962 as well as wars throughout the Middle East. But war in actual terms in the conflict between West as well as those of the Soviet Bloc never happened.

Germany was destroyed completely through invasions and bombs. The USSR established an undemocratic regime that ruled East Germany.

The most important German city of the time, Berlin which was the capital, was infested by a communist government. The enclave was a solitary democratic republic within the midst of a dictatorship communist was a huge issue for communists.

The USA as well as the UK as well as France were granted access to the right to access West Berlin. The Russians attempted to block from this by imposing a blockade of Berlin in 1948.

The blockade was removed through a massive Allied airlift. It was the beginning of the Cold War deepened, as China changed its mind to Communist after 1949.

The Russians were afraid that they would be annihilated if the USA might use their atom bomb against their country. The atom bomb ended the war with Japan to a close August 1945.

The Russians due to the spying of others and a huge effort purchased the atom bomb in the year 1949.

The end of the Second World War in Europe The Allies observed that Germans had a higher level of expertise in the field of rocketry in comparison to Britain or and the USA and the USSR.

The V1 rocket had destroyed parts of London in 1944 using the V1 rocket. They had then created a weapon with far better potential with V2 rocket. V2 rocket.

The Americans were extremely impressed, and were quick to take all from the German rocket programs into the USA as was possible.

Along with actual rockets they also took in all of the German rocket scientists as they could.

One of them named Wernher von Braun was a major player in the formation of what was later the US NASA space programme.

The mission to transfer the German rocket programme into the USA was dubbed Operation Paperclip. Sometimes it was believed that a blind eye was turned towards the war crimes that these individuals could have been guilty of.

Every combatant on the field had fired rockets, with incredible and sometimes destructive effect throughout the war.

There was a time when the British and Canadian armies utilized their Land Mattress rocket launcher in the last moments during World War II.

In the early days, United States army had little recourse to rocket launchers, except for a handful of rocket launchers that were mounted on tanks. However, their navy utilized them to amazing effect to attack Japanese posts during the Pacific War.

The Japanese employed them during later phases of the Pacific conflict in the Philippines, Iwo Jima, as well as Okinawa.

The Russians made use of their Katyusha rocket launcher that was referred to as the Stalin organ in the fight against Germans beginning in 1941 through the war.

The Germans also used a rocket launcher that is similar to Katyusha and frequently employed the device.

In the Second World War, the Germans created a powerful air bomb known as the V1. The V1 clearly came prior to the V2. Its impact was more severe for the British as compared to the V2.

The Germans unleashed hundreds of V1 bombs on London and caused lots of destruction as a result. V1 was the V1 was the forerunner of cruise missiles that were used during today's 21st Century.

As with like the V1 this cruise missile can not travel very fast, but is extremely precise and extremely destructive. Both Americans as well as Russia are using them to incredible effect in Syria.

The Americans believed that they had a great chance in combining massive rockets nuclear payloads, if they could develop the German V2 were to be further developed and further improved.

It was the reason they decided to shifting the German rocket project into the USA in the shortest time possible.

The Russians were also aware of this possibility and, as a result, was a constant advancement in rocketry on both sides. Both the USA as well as the USSR build what's known as ICBM rockets.

ICBM is an acronym for the term "intercontinental ballistic missile. They were created to be able to not just be fired from silos on land, but also from trains, and nuclear submarines.

It wasn't just the Americans as well as the Russians possess these weapons, but also the British as well as the French. They were upgraded so that a single missile could hold several nuclear warheads.

Missiles equipped with this ability were referred to as MIRVS. MIRV means multiple independently vehicles for reentry.

Another element in this Cold War was the relationship between Communist China as well as China's Soviet Bloc. The Chinese operated independently of other nations and obtained the nuclear bomb in 1964, and also an nuclear missile in the year 1966.

While the relationship between the West as well as that of the USSR with China is crucial during the Cold War it had little influence on the events that followed.

The other crucial aspect in ending Cold War was the complete fall of communism across the globe during the time between the late 1980s and the beginning of 1990s.

While of major importance, the event took place long following 1969, that year when the Moon landing, either alleged or not.

The book focuses on Apollo moon landings Apollo moon landing, and whether or not it happened, there are some who might wonder why there is a section dedicated to Cold War.

The Apollo project was quite expensive, and to have taken place without the Cold War.

This was because of the Cold War that provided the basis for what's known as"the Space Race between the USA and Russia.

If not due to this conflict in the West as well as the Soviet bloc, it's likely to think that Apollo missions could have ever taken place.

The Cold War kept on providing arguments for building of the Berlin Wall, the Cuban Missile Crisis as well as the South East Asian Wars for the Space Race.

Both America as well as Russia were able to see their successes on the space field as proof of the power of their technological capabilities. Both countries saw their participation in the Space Race as a way to show all the rest of humanity that their political technology was superior.

The final result in the Cold War was not decided through space race but rather by Space Race but by other aspects.

The story goes that the decision was made due to the fact that 'blue jeans' and Beatles was more attractive to young people in the Soviet Bloc than class war ethnic music, class wars and the promise of a glorious future which didn't materialize.

The next chapter will explain the story of Space Race history and describe the events that took place throughout Space Race. Space Race.

Chapter 14: The Space Race

Space rockets of today represent among mankind's numerous successes. They're the product of research and advancement of a multitude of years.

At present, rockets are the sole means for launching to space. As we enter the 2nd decade of the 21st century there are efforts underway to alter this which you can read about in the final chapter.

How long ago were the first rockets? In the year 100 A.D., the Chinese had gunspowder. This was used to fuel the first rockets.

It's unclear if the initial Chinese experimenters in rocketry knew the impact their inventions would have on the world for hundreds of thousands of years in the future.

The initial use for military rockets was made by Chinese who put these missiles onto arrows that were shot by archers.

It didn't take too much time to realize that no need for arrows and rockets were already on the course. They were not widely used and was not guaranteed at this time.

The beginning of the thirteenth century there was war between the Mongols as well as the Chinese. In one fight, they Mongol lost to Chinese who employed a number of the basic rockets.

For those who study rocketry as a science, it was an opportunity to show their capabilities, but there is a doubt that participants in that older combat realized that.

Much as is often when it comes to weapon systems, Mongols created rockets to use for themselves and it's likely the Mongols who Europeans first learned about rockets and the potential of them to be used as weapons.

Over the next 200 years, advances of rocketry were observed across a variety of European countries, which included Italy, France, and England.

Newton developed the Three Laws of Motion in the last quarter of the 17th century. The laws of Newton fully explained the force and movement in conditions that were slow in speed.

They were the science-based basis of rocketry today. By following these rules the rockets could be created with a methodical approach.

There was a lot of development happening across Europe to use during conflict. These were employed to tremendous effectiveness during the war of British in the fight against United States in the 1812 conflict. They were able to achieve great successes in destroying American defenses.

until the close in the First World War, there was no significant role for rockets during the bloody fights of that terrible battle, as artillery with high power proved to be more efficient in damaging opponents.

They first saw use by the French in the war of Verdun as part of an air to ground mission. The British utilized them against German airships, but they weren't particularly effective.

In Russia in the year 1898 when the man named Tsiolkovsky with a great sense of foresight, came up with the idea of exploring space using rockets.

In the following years, it was he that suggested using liquid fuel to fuel rockets. If this was used, to propel the rocket, the distance it traveled could be considerably enhanced.

The early 20th century There was an outstanding American called Robert Goddard who had a huge fascination with rockets. He was a prolific researcher on the rockets, and because of his work, abilities of rockets significantly enhanced.

The Second World War saw rockets become an effective weapon. It was during this time that the Germans received a huge shock during 1941, with they discovered that the Katyusha rocket-powered system, also known as the Stalin organ was employed by the Russians during the Battle of Moscow.

They proved to be powerful weapons. The Germans are so impressed the Germans later developed and utilized similar weaponry.

In the island-hopping wars of the Pacific War, the Americans employed rockets similar to those that were launched from land craft.

with much more powerful rockets, and a much larger arsenal of rockets, the Germans far outpaced the Allies and constructed a strong rocket, the V2. Sadly for them the V2 was not built until to be too late to stop their loss.

Wernher von Braun, who was the head of the team that created these, later went on to manage the space program of the Americans between 1950 and 1960s.

After World War II ended the Cold War began. The two superpowers fighting one another.

In the 1950s and beyond, space was a new area for this raging competition. Every country tried to demonstrate that its technological capabilities or military capability, as well as the way it operated its economy were superior.

In the mid-point in the 1950s, Cold War was a major factor in the life of most nations. The reason for this was an arms race as well as an ever-growing threat from atomic weapons as well as by the spying between two different systems.

Tensions like these were commonplace during the time period between 1946 and 1989. The tensions were further aggravated through events such as the fall of the Berlin wall of 1961 as well as in 1962, the Cuban missile crisis of 1962, and also in the Vietnam War.

On the 4th of October of 1957 on the 4th of October 1957, an Russian missile was launched Sputnik as the world's first synthetic satellite as well as the first artificial object that humans to be placed in the earth's orbit. It was the day that marked the beginning of what's known as the Space Race.

The top leadership in America USA was shocked and made a number of changes to the system of education. One of the changes included an idea known as "the New Math.

The New Math was a peculiar concept that was promoted by a few educationalists. The idea was that if primary students and secondary school pupils learned abstract algebra as well as more advanced concepts, they could be prepared in the future to be competitive with the Russians.

The idea was adopted in the following decades across the Western world, and ended up being complete catastrophe.

The explosion of Sputnik was quite a frightful unexpected surprise for the

Americans. They realized that rockets like such as R-7 the one that been launched by Sputnik may be utilized to fire nuclear warheads at the USA as well as her allies.

It was crucial for the Americans didn't fall way too much ahead of the Russians.

In the year 1958, America began the launch of the first satellite. It was named Explorer 1. It was developed in the US Army under the direction by the German NASA scientist Wernher von Braun.

During 1958 President Eisenhower set up NASA, the National Aeronautics and Space Administration.

The president Eisenhower created an approach by that orbiting satellites could be utilized for intelligence to determine what Russians as well as their allies were up to.

In 1959, the Russians achieved landing their first car on the moon. The vehicle was named Lunar 2.

In the month of April, 1961 Russian astronaut Yuri Gagarin was the first person to orbit Earth using an orbital spacecraft, which was named Vostok 1.

It wasn't until 1961 when the Americans could put an astronaut into space. The man was named Alan Shephard, who ventured to space on the 5th of May in 1961, even though the spacecraft did not reach space, made this.

Then in May 1961 in May 1961, the president of the United States, Kennedy declared his intention that United States would land a human on the moon before the end of this decade. The plan to construct the lunar landing platform was dubbed Project Apollo.

It was 1962 when John Glenn, who eventually was a senator before dying in December of 2016, was one of the very first American to be the first person to orbit the Earth.

www.ingramcontent.com/pod-product-compliance
Lightning Source LLC
Chambersburg PA
CBHW070556010526
44118CB00012B/1340